SpringerBriefs for Data Scientists and Innovators

Volume 2

Editor-in-Chief

Osamu Sudoh, Graduate School of Interdisciplinary Informatics, The University of Tokyo, Tokyo, Japan

This series presents concise summaries of cutting-edge research and practical applications in the area of big data analysis, decision making and prediction. With "innovation" as a key word, the series aims for sharing new approaches and inspiring ideas of data analysis arising from different fields including social sciences, artificial intelligence, medical care, security, policymaking, urban planning, and more. The series aims to promote data sciences for new innovations such as the use of big data to find cutting-edge solutions for human society at large.

Featuring compact volumes of 50 to 125 pages (approximately 20,000-45,000 words), Briefs allow authors to present their ideas and readers to absorb them with minimal time investment.

Typical texts for publication might include:

- A snapshot review of the current state of a hot or emerging field
- A concise introduction to core concepts that students must understand in order to make independent contributions
- An extended research report giving more details and discussion than is possible in a conventional journal article
- A manual describing underlying principles and best practices for an experimental technique

The standard concise author contracts guarantee that:

- an individual ISBN is assigned to each manuscript
- each manuscript is copyrighted in the name of the author
- the author retains the right to post the pre-publication version on his/her website or that of his/her institution

The publication of all volumes in SBDSI is to be done by a peer-reviewed process.

More information about this series at https://link.springer.com/bookseries/15750

Naoki Nakashima

Editor

Epidemiologic Research on Real-World Medical Data in Japan

Volume 2

 Springer

Editor
Naoki Nakashima
Kyushu University Hospital
Fukuoka, Japan

ISSN 2520-1913   ISSN 2520-1921   (electronic)
SpringerBriefs for Data Scientists and Innovators
ISBN 978-981-19-1621-2   ISBN 978-981-19-1622-9   (eBook)
https://doi.org/10.1007/978-981-19-1622-9

This Springer imprint is published by the registered company Springer Nature Singapore Pte Ltd.
The registered company address is: 152 Beach Road, #21-01/04 Gateway East, Singapore 189721,
Singapore

Preface

This book is devoted to understanding how medical big-data projects are developing in Japan.

Japan is the first country in global history to experience an aging society. Labor productivity has decreased accordingly without innovations to resolve this issue. Big-data analysis by Japanese medical Real-World Database (RWD) is one of the candidates for innovation to tackle the aforementioned issue.

First, this book discusses the original Japanese system that generates medical RWDs, in the hospital medical records system, the nationwide standardized health checkup system, and the public medical insurance system in Japan to establish background knowledge for Japanese medical big-data analysis. The exhaustive data of 120 million citizens possibly constitutes one of the largest medical data sets available worldwide, and analyzing these data is fascinating for data scientists, industries, and the public all over the world.

Next, the book introduces four representative big-data projects in the healthcare-medical field in Japan. Each project utilizes different data characteristics, but all projects are expected to be effective in changing the future of Japan. Readers can understand big pictures and concrete outcomes of the projects by themselves.

Then, this book also explains the importance of creating information standards to maintain data quality and to analyze medical big data. Readers can self-analyze which standards are installed in which RWDs, how the standards are maintained, and what types of issues are prevalent in Japan.

Pathology (phenotype) from RWD is extracted by the phenotyping method based on certain rules. This method is important for medical RWD analysis in any country and should be developed in each country because the method of analysis varies with each country. We hope this book contributes to establishing phenotyping rules in other countries.

In observational studies involving the secondary use of RWDs, researchers or data scientists should consider special aspects of the analysis method. To improve the quality of big-data analysis, study design should be emphasized as an important part of the startup phase of research.

Finally, this book describes the ethical process involved in big-data projects of medical RWDs in Japan. Regarding this issue, this book explains the "Next Generation Medical Infrastructure Act", which was enforced in 2018 and will promote medical science and industries in Japan by utilizing data from medical RWD in Japan.

Readers can analyze the following from the book:

1. The mechanisms involved in the generation of Japanese medical RWDs and the aspects or characteristics of the data. Subsequently, readers can understand how Japanese big-data analysis works in Japanese society.
2. The four representative big-data projects in Japan on a national scale, including each purpose, method, aspect, situation, and issue.
3. Basic technology issues (including ethics) and solutions in conducting Japanese medical RWD analyses.

I appreciate the great efforts that the authors have made to publish this book in this critical state of COVID-19 pandemic.

Finally, my special thanks goes to Ms. Masako Ito for good management of all stages of the publishing process.

Fukuoka, Japan   Naoki Nakashima
July 2021

Overview

This book is devoted to developing an understanding of how medical big-data projects have developed in Japan.

Japan was the first country in world history to experience an aging society. Medical costs have drastically increased, and labor productivity has fallen accordingly, in the absence of any innovation that could resolve these issues. Big-data analysis using medical Real-World Data (RWD) on Japan may be a possible means for innovation aimed at tackling these issues.

The history of Japanese medical system, policy, and function are examined here so that foreign researchers can more easily understand them. Additionally, this book is also intended to provide a reference for Japanese researchers to be able to use to indicate their research methodology in their original articles so as to be able to avoid describing their practice at length in the context of a restricted word count. This book consists of a total of nine parts in Vol. I and Vol. II together, written by outstanding Japanese researchers. In this overview, I give an outline indicating each part.

Japanese Health Care and the Medical Information System

Japanese Public Medical Insurance System

First, the Japanese medical system, which generates RWD in the hospital Electronic Medical Records (EMR) systems, the Nationwide Standardized Health Checkup System (NSHC), and the public medical insurance system are examined to establish the background information necessary to conduct big-data analysis of the Japanese medical system. This set of exhaustive data on 126 million citizens is possibly one of the largest medical data sets available in the world, and these data are a treasure for data scientists, industries, and the public all over the world due to Japan's long history of a public medical insurance system covering all Japanese citizens.

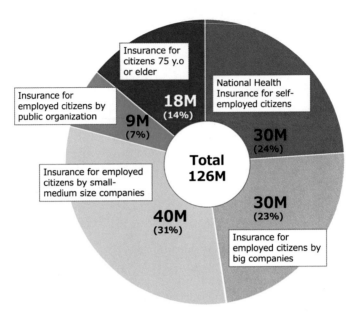

Fig. 1 Categories of Japanese public insurance system according to share of population insured

Japan's public medical insurance system was launched in 1961. This system covered all citizens and allowed free access to any medical institution; it was cutting edge at the time, but it stills the two features mentioned as universal insurance and free access.

However, there are many insurers in business in Japan (3,403 in 2018) to cover all Japanese citizens, although this total is less than the original number that had been in 1961. Japanese insurers can be classified into three categories: insurance for the self-employed (1,878 insurers in 2018, matching the number of local governments, *i.e.*, cities, towns, and villages), employee insurance (1,478 insurers in 2018), and insurance for those 75 years old or older (47 insurers in 2018, matching the number of prefectural governments) [1]. At the end of March 2018, on a rough numerical basis, insurance for the self-employed covers 30 million people (24% of all citizens), employee insurance covers 78 million (62%), and insurance for those who are 75 years old or older covers 18 million (14%), reaching nearly the total number of citizens (126 million) (see Fig. 1) [1].

Patient payment rates for medical expenses have been altered many times. At the present rate (in 2019), citizens between 6 and 69 years old pay 30% of the cost, those under 6 years old or those between 70 and 74 years old pay 20%, and those over the age of 75 years pay 10% of the cost (however, if a 70-year-old or older person has an income comparable to the younger generation, he or she must pay 30% of the cost). For example, if a 50-year-old is receiving services, the provider collects 30% of the medical fee from the patient and 70% from the insurer.

In Japan, medical services are provided by 8,324 hospitals (defined as medical institutions with 20 or more beds) and 102,396 clinics (medical institutions with less than 20 beds), 68,488 dental clinics as of May 2019 [2]. Dispensing pharmacies with number of 60,171 are also working as of March 2020 [3].

In these institutions, 327,210 medical doctors; 104,908 dentists; 311,289 pharmacists; 1,218,606 nurses; 52,955 public health nurses; and 36,911 certified midwives were working in December 2018 [4, 5].

Japanese medical costs have been increasing with the aging of its population. The costs reached JPY 43.6 trillion in 2019 [6], and this amount may continue to increase. This means that increasing the cost-effectiveness of medical services is an urgent national task.

Digitalization and Reuse of Insurance Claim Data and Health Checkup Data

Insurance claims were processed on paper, from 1961 until the beginning of the twentieth century, so that the data produced were little used or analyzed. Insurance claim data began to be digitized at the beginning of the twenty-first century, reaching up to a level of 99%, that is, almost complete digitization, at present (2021). All medical procedures covered by insurance are recorded in the claims data, allowing the insurers to evaluate what procedures were provided to which insured person by the medical institutions they visited.

In 2003, a new payment system, called Diagnosis Procedure Combination (DPC) was put in place in all of the 82 university hospitals in Japan. DPC is a Japanese variant of the familiar Diagnosis Related Group/Prospective Payment System used in the United States. DPC is now applied in 1,730 hospitals (in almost all acute care hospitals) in Japan.

In 2008, the NSHC was implemented as a duty of the insurer to detect and prevent Non-communicable Diseases (NCDs), such as diabetes mellitus, hyperlipidemia, dyslipidemia, or chronic kidney diseases (CKDs). The target age group for NSHC is 40- to 74-year-olds (54 million people in 2019). The number of NSHC examinees has been increased, from 38% in 2008 during the implementation to 57% in 2019 [7]. The results of NSHC are shared with both the examinee (the insured) and the insurer.

In 2015, the Ministry of Health, Labour and Welfare (MHLW) initiated a promotional plan called the DataHealth (DH) plan, against the background of the progression of the digitalization of insurance claims and NSHC. The DH plan coordinates the population management process for the insured by the insurers, matching personal claims data and NSHC data. In this program, insurers provide personal support services on the basis of the results of matching in order to reduce their risks.

For example, if an insurer does not have both claim data and NSHC data, it should recommend the given insured person to undergo NSHC screening because

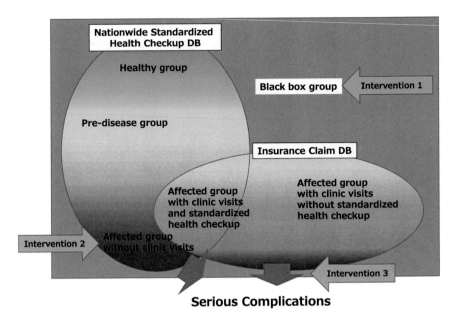

Fig. 2 Three important intervention points in the DH plan

the insurer does not know the individual's precise health condition (intervention point 1 in Fig. 2). If the insurer does not have claims data but can obtain the results of an NSHC screening that shows hints of an NCD, the insurer should urge the insured to visit a clinic or hospital because this fact would imply that the insured had ignored the results of an NSHC screening (intervention point 2 in Fig. 2). Furthermore, if the insurer is able to obtain claims data reliably, but the results of the annual NSHC screening show no improvement or indeed are worse than those of the previous year, the insurer should intervene strongly through public care nurses to prevent serious complications from developing (intervention point 3 in Fig. 2). As noted above, MHLW began promoting nationwide population management using disease management methodologies for primary to tertiary prevention of NCD through the digitalization of insurance claim data and NSHC by public medical insurers soon after the turn of the twenty-first century. Since 2018, MHLW has been using a new carrot and stick policy called Support System for Insurer Efforts. With this system, MHLW imposes an economic penalty to insurers that have insufficient outcomes for their support of insured individuals with high risks. It is also planned that MHLW will provide an economic incentive for insurers with better outcomes from 2020.

Japanese Electronic Medical Records

In the analysis of medical RWD, Electronic Health Record (EHR) data, which includes clinical outcome data such as lab tests and diagnosis data, are more useful than insurance claim data, which only shows the clinical procedures. However, EHR is not yet widely integrated in Japan; they remain distributed in several medical institutions as Electronic Medical Records (EMR). EMR is limited to data from only those who had visited the given institution. The NSHC database includes healthy subjects, but these data are held separately by the 3,418 insurers and are not integrated, except in the form of an anonymous National Data Base (NDB), only available for secondary use. More on the NDB will appear below.

All told, the digitization of medical records in Japan in 2017 was only 46.7% for hospitals and 41.6 % for clinics, although the number has gradually been increasing.

There has been a recent tendency to establish standard data repositories outside of EMR for data for primary or secondary use by multiple institutions because it is too late and too difficult to standardize existing EMR. The standard Standardized Structured Medical Information eXchange version 2(SS-MIX2), is listed as an MHLW standard. See Part IV in Vol. I) is used widely in Japan. At the end of March 2018, 1,360 hospitals had already installed SS-MIX2. It must be considered how to establish an integrated EHR system that can last a person's lifetime and follow that person, and SS-MIX2 is a powerful candidate for the skeleton of such a system of Japanese EHR (see Part IV in Vol. I).

The idea of a Personal Health Record (PHR) is expected to gain greater currency in the near future and to become the main data sources for EHR. As the smartphone gains greater and greater penetration, the PHR is expected to be a valuable tool for patient engagement. However, ideas around PHR are still in the research stage, and the PHR service model has not yet spread to Japan. Therefore, this book does not discuss it in detail.

Since 2018, six clinical associations in Japan: the Japan Diabetes Society (JDS), the Japan Association for Medical Informatics (JAMI), the Japanese Society of Hypertension, the Japan Atherosclerosis Society, the Japanese Society of Nephrology (JSN), and the Japanese Society of Laboratory Medicine, have recommended a configuration for PHR using standardized data item sets [8, 9]), therefore, PHR can thus be expected to be seen as a tool of patient engagement and a powerful data source for EHR in Japan in the near future.

Japanese Big-Data Projects Based on RWD

Next, we introduce three representative big-data projects in the Japanese fields of healthcare and medicine. Each project features different data characteristics, but each project is expected to be effective in changing Japan's future. Readers can come to understand the big picture and develop the concrete outcomes of the projects by themselves (see Part IV in Vol. II).

Diagnosis Procedure Combination (Outline of Part I in Vol. I)

The first nationwide Data-Driven Medical Study (DDMS) in Japan was the DPC.

The DPC, which was launched in 82 hospitals (mainly university hospitals) in 2003, was designed not only for use as a payment system, but also for the creation of high-quality data sets to be used in the analysis. Because diagnostic data directly decide payment of medical fees for inpatients in the DPC, the hospital is not able to upcode diagnoses (this would be fraud), which results in the registration of a precise diagnosis for each admission. Although the DPC was only used for inpatients, and some diseases were excluded (for example, psychiatric diseases), it eventually spread to 1,730 hospitals, including almost all acute care hospitals, and 7 million cases were registered in 2017. The DPC has already contributed to quality assessment and standardization in Japan's acute care hospitals (see Part I in Vol. I).

National Database of Health Insurance Claims and Specific Health Checkups of Japan (Outline of Part II in Vol. I)

Each insurer must provide all claims data and NSHC data to MHLW. Then, MHLW matches and anonymizes them to establish the National Database of Health Insurance Claims and Specific Health Checkups of Japan (NDB). The NDB is only used for research purposes on an application–examination basis.

The NDB features one of the world's largest data sets, including all claim data matched with all NSHC data from the 126 million citizens of Japan. Although it has limitations (for example, it excludes those living on welfare—about 2 million citizens), and it excludes disqualification data, such as death events. The NDB excludes outcome data, such as lab test result data, except for NSHC data. In all, it yields 2 billion cases and 40 billion data points each year.

NDB has responded to requests from researchers since 2013 and helped produce research outcomes by extracting partial datasets. Recently, Dr. Naohiro Mitsutake of the Institute for Health Economics and Policy established an analysis infrastructure using all claim data (10 billion cases and 200 billion data points over 6 years, 2009–2014), using an ultra-high-speed search engine developed by Professor Masaru Kitsuregawa at Tokyo University. Using this infrastructure, for example, the author, at Kyushu University, is conducting an analysis of NDB while Professor Yamagata and his colleagues at Tsukuba University are studying CKD and end-stage kidney diseases (see Part II in Vol. I).

Medical Information Database Network (Outline of Part III in Vol. I)

The Medical Information Database Network (MID-NET) project is being implemented by the MHLW and the Pharmaceuticals and Medical Devices Agency to detect adverse drug events after marketing approval, using a pharmaco-epidemiology method, together. In other words, this is a Japanese version of the Sentinel initiative that is underway in the United States.

In all, 23 hospitals part of 10 institutions (8 are university hospitals) have joined the project, providing real-time data for 4 million patients (diagnosis, results of lab tests, prescriptions, and claims data) through SS-MIX2 and the claims database. The MID-NET project spent more than five years on validating the data, and it can now boast of the data quality. The project was formally launched in 2018, after seven years of preparations, to provide a system for use by medical researchers and pharmaceutical companies. The results are expected to include detection of serious adverse events of drugs, even if late, in the so-called long tail of cases (see Part III in Vol. I).

Disease Registration Cohort Study with EDC from EMR (Outline of Part IV in Vol. I)

AMED (Japan Agency for Medical Research and Development), an important Japanese provider of medical grants since 2015, has activated a DDMS (including medical image analysis) and prospective disease registration research project conducted by clinical academic associations. This project is collecting and analyzing RWD from SS-MIX2, the claims database, and standard DICOM image data from clinical image databases connected to picture archiving and communication systems from multiple medical institutions.

The J-CKD-DB project is a representative DDMS project that is being conducted by multiple medical institutions. It only uses clinical RWD, without adding any manual data input for research purposes to avoid burdening clinical sites. J-CKD-DB is being led by JSN and has already collected more than 100 thousand cases of CKD (see Part IV in Vol. I).

The medical image DDMS coordinated by AMED includes radiology, pathology, gastrointestinal endoscopy, retina examination, and echogram images. These are to be used to develop AI that can support the interpretation of medical images by doctors.

The Clinical Core Hospital Research Network project, launched by the MHLW as an AMED project in 2018, is intended to establish infrastructure for DDMS among 12 clinical core hospitals.

Some research projects are using EMR for their data source, although they are aiming at prospective clinical research. For example, the template function in EMR is available for inputting clinical data, which are usually described with free text

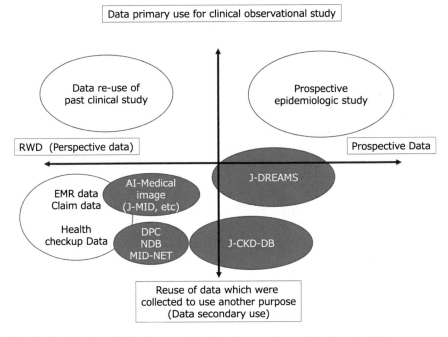

Fig. 3 Classification of clinical observational studies according to data characteristics

in EMR. Medical doctors can use template functions in their daily clinical work. J-DREAMS (acronym for Japan Diabetes compREhensive database project based on an Advanced Electronic Medical Record System), which is being conducted by the National Center for Global Health and Medicine and JDS, is a typical example of this pattern (see Part IV in Vol. I). For these prospective epidemiologic studies or registration cohort studies, SS-MIX2 is expected to be used as the source of data for EMR in dominant cases. Figure 3 shows the relations among different types of clinical observational study.

Importance of Data Quality in Big-Data Analysis

Clinical Pathways (Outline of Part I in Vol. II)

After Nurse Karen Zander developed the clinical pathway in the 1980s, some Japanese hospitals put it in place for simple planning or for scheduling of medical service. At the end of the twentieth century, the Japan Society of Clinical Pathway (JSCP) began promoting the diffusion of outcome-oriented clinical pathways and continuing up to the present.

In Japan, more than 2,000 hospitals (largely acute care hospitals), out of a total of 8,442, are already using clinical pathways. Using these clinical pathways, clinical outcome data are easily obtained from EMR as structured data, although such data is often described in free text as progress notes. Many EMR packages already contain the clinical pathway function already, however, there is no standardization among vendors at present. Therefore, the analysis of clinical pathways with multiple medical institutions can be difficult.

Thus, in 2016, the JSCP and JAMI established a collaborative committee to standardize the clinical pathway system, and they initiated a model project called the ePath Project with four hospitals and four top EMR vendors in October 2018 until March 2021, funded by AMED (see Part I in Vol. II).

Standard Codes (Outline of Part II in Vol. II)

This book also explores the importance of creating information standards to maintain data quality and to help analyze medical big data.

The largest general issue for RWD is data quality. Before the analysis of RWD for secondary use, data cleansing (correction of duplicate data, location of missing or mixed data, assessment of outliers, and so on) must be undertaken. Furthermore, the MID-NET project found that data mapping to standard code is moving slowly or has even been incorrectly done, even in institutions that are using SS-MIX2. We expect that researchers and medical institutions will come to recognize the importance of data quality and try to increase it by conducting big-data analysis of medical data.

However, we cannot expect that the data quality of DDMS will be perfect even after data management. It is also very important to develop methods to evaluate the data quality with exactness. For this book, Professor Dongchon Kang of Kyushu University provided an overview of lab tests mapping between local codes and the standard JLAC10 codes, as well as to JLAC11, the next generation version of JLAC10. The ICD10 and ICD11 standard codes for diagnosis are introduced, as well as HOT, the drug standard code (see Part II in Vol. II).

We also introduce a central management trial for the standard code (ICD10, JLAC10, and HOT) in multiple medical institutions (in other words, we review the governance of standard codes) in the aforementioned MID-NET project with AMED funding (see Part II in Vol. II). Readers can analyze which standards should be installed in which RWDs, how the standards should be maintained, and what types of issues are prevalent in Japan.

Data Quality and Phenotyping (Outline of Part III in Vol. II)

RWD often lacks data items for its analysis that would be critical for achieving goals. Japanese EMR and claims data do not have clinically appropriate phenotypes as pathologies or diagnoses in databases. What does exist is the only diagnosis for insurance reimbursement, and this is not a medically correct phenotype. Therefore, a methodology should be developed, called ePhenotyping, to detect or predict correct phenotypes from other data items (see Part III in Vol. II).

ePhenotyping would be very important for medical RWD analysis in any country, and it should be developed for each one because the medical system is different in each. We hope that this book can contribute to the establishment of ePhenotyping rules in Japan and other countries as well.

Data Analysis of Real-World Data (Outline of Part IV in Vol. II)

In observational studies involving the secondary use of RWDs, researchers and data scientists should consider the special aspects of their analysis methods. To improve the quality of their big-data analysis, their study designs should be emphasized during the initiation phase of research.

For big-data analysis, adding to the knowledge of found by classical biomedical statistics, new analytical methodologies such as machine learning, that previously were not used for medical analysis, should be incorporated. Furthermore, it is impossible to design appropriate research plans without knowledge of the system and local operations in the clinical sites, where RWD accumulate. Of course, it would be difficult for one person to have all of this knowledge, but at least one member on the project team should have expertise in each needed area, such that the team as a whole can cover all necessary knowledge to ensure a proper big-data analysis and produce papers for publication on a high level (see Part IV in Vol. II).

Ethical and Other Issues of Data Regarding Secondary Data Use in Japan (Outline of Part V in Vol. II)

Finally, this book describes the ethics of big-data projects using medical RWD. Before we begin clinical observational research projects, the revised Personal Information Protection Act of May 2017, should be understood, along with the simultaneously revised ethical guidelines for medical research. This book also explains the Next-Generation Medical Infrastructure Act, which came into law in May 2018 and promotes medical science and industries in Japan by allowing the use of medical RWD in Japan. Of course, regulations should be complied with, but the benefit for patients and citizens should be uppermost (see Part V in Vol. II).

Recently, the internet of things has extended a wide influence into daily life, and this implies that many devices in our lives are creating and accumulating digital data on our daily vital signs and behavior. The processing and analysis of the explosively increasing amounts of health and medical data yielded by these remains an issue. Further, the increasing amounts of genetic information are producing the same effect. However, the present numbers of Japanese data scientists are apparently insufficient to tackle these serious issues, relative to the numbers in other countries. We should make haste to increase their numbers.

Naoki Nakashima, M.D. Ph.D.
President of Japan Association of Medical Informatics
Director/Professor
Medical Information Center
Kyushu University Hospital
Fukuoka, Japan
nakashima.naoki.351@m.kyushu-u.ac.jp

References

1. MHLW: 22nd Medico-economical fact-finding surveillance. https://www.e-stat.go.jp/stat-search/files?page=1&toukei=00450392&result_page=1. (article in Japanese, Retrieved on Aug 7, 2021)
2. MHLW: Dynamic Surveillance of Number of Medical Institutes. https://www.mhlw.go.jp/toukei/saikin/hw/iryosd/m19/dl/is1905_01.pdf. (article in Japanese, Retrieved on Aug 7, 2021)
3. MHLW: Dynamic Surveillance of Number of Dispensing Pharmacies. https://www.mhlw.go.jp/toukei/saikin/hw/eisei_houkoku/19/dl/kekka5.pdf. (article in Japanese, Retrieved on Aug 7, 2021)
4. MHLW: Dynamic Surveillance of Numbers of Doctor, Dentist and Pharmacist. https://www.mhlw.go.jp/toukei/saikin/hw/ishi/18/dl/kekka.pdf. (article in Japanese, Retrieved on Aug 7, 2021)
5. MHLW: Dynamic Surveillance of Numbers of Nurse, Public Nurse and Midwife. https://www.mhlw.go.jp/toukei/saikin/hw/eisei/18/dl/kekka1.pdf. (article in Japanese, Retrieved on Aug 7, 2021)
6. MHLW: Dynamic Surveillance of Medical Cost. https://www.mhlw.go.jp/stf/newpage_13214.html. (article in Japanese, Retrieved on Aug 7, 2021)
7. MHLW: Dynamic Surveillance of Nationwide Standardized Health Checkup. https://www.mhlw.go.jp/content/12400000/000755573.pdf. (article in Japanese, Retrieved on Aug 7, 2021)
8. Nakashima N, et al. (2019) Journal of Diabetes Investigation, 10 (3): 868–875
9. Nakashima N, et al. (2019) Diabetology International, 10 (2): 85–92

Contents

Contributors

Tatsuo Hiramatsu Department of Medical Informatics, International University of Health and Welfare, Tokyo, Japan

Takeshi Imai Center for Disease Biology and Integrative Medicine, Graduate School of Medicine, The University of Tokyo, Tokyo, Japan

Rieko Izukura Medical Information Center, Kyushu University Hospital, Fukuoka, Japan

Dongchon Kang Department of Clinical Chemistry and Laboratory Medicine, Kyushu University Graduate School, Fukuoka, Japan

Yutaka Matsuyama Department of Biostatistics, School of Public Health, The University of Tokyo, Tokyo, Japan

Naoki Nakashima Medical Information Center, Kyushu University Hospital, Fukuoka, Japan

Mihoko Okada Institute of Health Data Infrastructure for All, Tokyo, Japan

Jinsang Park Department of Pharmaceutical Sciences, School of Pharmacy at Fukuoka, International University of Health and Welfare, Fukuoka, Japan

Tomohiro Shinozaki Department of Information and Computer Technology, Faculty of Engineering, Tokyo University of Science, Tokyo, Japan

Hidehisa Soejima Saiseikai Kumamoto Hospital, Kumamoto City, Japan

Atsushi Takada Medical Information Center, Kyushu University, Fukuoka, Japan

Shoji Tokunaga Medical Information Center, Kyushu University Hospital, Fukuoka, Japan

Yoshifumi Wakata Medical Information Management Center, National Hospital Organization Kyushu Medical Center, Fukuoka, Japan

Ryuichi Yamamoto Medical Information System Development Center, Tokyo, Japan

Abbreviations

ACD	Administrative Claims Data
AI	Artificial Intelligence
AMED	Japan Agency for Medical Research and Development
AUC	Area Under the Curve
CDS	Clinical Decision Support
CKD	Chronic Kidney Diseases
CP	Clinical Pathways
CSV	Computerized System Validation
CT	Computed Tomography
DDMS	Data Driven Medical Study
DICOM	Digital Imaging and COmmunications in Medicine
DIR	Dose Index Registry
DPC	Diagnosis Procedure Combination
DRG	Diagnosis Related Group
DRG/PPS	Diagnosis Related Group/Prospective Payment System
EDC	Electronic Data Capture
eGFR	estimated Glomerular Filtration Rate
EHR	Electronic Health Record
EMR	Electronic Medical Records
ESKD	End-Stage Kidney Disease
ETL	Extract/Transform/Load
GBDT	Gradient Boosting Decision Tree
GFR	Glomerular Filtration Rate
GPSP	Good Post-Marketing Study Practice
HELICS	Health Information and Communication Standard Board
HIS	Hospital Information System
HL7 CDA R2	HL7 Clinical Document Architecture Release 2
ICD10	International Statistical Classification of Diseases and Related Health Problems, 10th revision
ICD11	International Statistical Classification of Diseases and Related Health Problems, 11th revision

ICMRA	International Coalition of Medicines Regulatory Authorities
ICT	Information and Communication Technology
IHE	Integrating the Healthcare Enterprise
IHEP	Institute for Health Economics and Policy
IHE PDI	IHE Portable Data for Images
IHE XDS	IHE Cross-provider Document Sharing
IRB	Institutional Review Board
IVD	In Vitro Diagnostic
JADER	Japanese Adverse Drug Event Report
JAHIS	Japanese Association of Healthcare Information Systems Industry
JAMI	Japan Association for Medical Informatics
JCVSD	Nationwide Japan Adult Cardiovascular Surgery Database
J-DREAMS	Japan Diabetes compREhensive database project based on an Advanced electronic Medical record System
JDS	The Japan Diabetes Society
JIRA	The Japan Medical Imaging and Radiological Systems Industries Association
JLAC10	Japanese Laboratory Codes, Version 10
JMDC	Japan Medical Data Center
J-MID	Japan Medical Imaging Database
J-QIBA	Japan's Quantitative Imaging Biomarker Alliance
J-RIME	Japan Network for Research and Information on Medical Exposures
JRS	Japan Radiological Society
JSCP	Japanese Society for Clinical Pathway
JSDT	Japanese Society for Dialysis Therapy
JSLM	Japan Society of Laboratory Medicine
KT	Kidney Transplantation
LOINC	Logical Observation Identifiers Names and Codes
MCDRS	Multi-purpose Clinical Data Repository System
MDC	Major Diagnosis Category
MDV	Medical Data Vision
MERIT-9	MEdical Record, Image, Text-Information eXchange
MHLW	Ministry of Health, Labour and Welfare
MID-NET	Medical Information Database NETwork
MIHARI	Medical Information for Risk Assessment Initiative
MRI	Magnetic Resonance Imaging
NCD	National Clinical Database
NCD	Non-Communicable Disease
NCGM	National Center for Global Health and Medicine
NDB	National Database of Health Insurance Claims and Specific Health Checkups of Japan
NPU	Nomenclature for Properties and Units
NRS	Numerical Rating Scale
NYHA	New York Heart Association
OHDSI	Observational Health Data Sciences and Informatics

OMOP	Observational Medical Outcomes Partnership
PACS	Picture Archiving and Communication Systems
PHR	Personal Health Record
PMDA	Pharmaceuticals and Medical Devices Agency
PPV	Positive Predictive Value
RCT	Randomized Controlled Trial
RECORD	REporting of studies Conducted using Observational Routinely-collected health Data
RWMD	Real-World Medical Data
SAERs	Spontaneous Adverse Event Reports
SDMT	Standard Diabetes Management Template
SDV	Source Document Verification
SOAP	Subject, Object, Assessment and Plan
SS-MIX2	Standardized Structured Medical Information eXchange 2
STROBE	STrengthening the Reporting of OBservational Studies in Epidemiology
VPN	Virtual Private Network
XCA	Cross-Community Access

List of Figures

Data Quality Governance Experience at the MID-NET Project

Phenotyping in Japan

Integration of Phenotyping Algorithms in Japan

The Next-Generation Medical Infrastructure Law

List of Tables

Clinical Pathway

Real World Medical Data and Clinical Pathway in Japan

Hidehisa Soejima

1 Clinical Pathway in Japan

Clinical pathways (CP) were developed by Karen Zander et al. in the 1980s [1]. CP were epoch-making in the sense that traditional critical pathways used in engineering were applied to medical treatment. Zander put the care map on the market commercially, but the concept of CP has not been organized systematically and academically and as such is not yet a useful tool. In fact, the introduction of DRG triggered the spread of CP in the United States, and it seemed that the desire for more efficient medical treatment and schedule management were significant. Presently, it can be argued that since the protocol is used as a schedule, it serves only a similar function to schedule management.

Of course, in the process of variance analysis, Zander created an outcome-oriented CP with the main objectives of process management and goal management, but it has yet to be widely adopted.

On the other hand, in Japan, CP studies began in the early 1990s. In the mid-1990s, CP was initially introduced mainly in acute hospitals. Originally, it was a schedule-like pathway where was no target management concept. However, in the 2000s goals were clearly defined, and medical professionals began to support an outcome-oriented CP, which lead to its spread. The term variance is used to describe cases where the outcome was not achieved. By analyzing non-standard events that may affect the success or failure of treatment, medical professionals can expect to improve treatment outcomes. In the past, paper charts were prevalent and it was difficult to obtain large amounts of data, but using the technique called pathway to visualize medical processes, manage goals, and improve treatment attracted the attention of many medical staff in 2000, and the Japanese Society for Clinical Pathway

H. Soejima (✉)
Saiseikai Kumamoto Hospital, 5-3-1 Chikami, Minami-ku, Kumamoto City 861-4193, Japan
e-mail: hidehisa-soejima@saiseikaikumamoto.jp

© The Author(s), under exclusive license to Springer Nature Singapore Pte Ltd. 2022   3
N. Nakashima (ed.), *Epidemiologic Research on Real-World Medical Data in Japan*,
SpringerBriefs for Data Scientists and Innovators 2,
https://doi.org/10.1007/978-981-19-1622-9_1

(JSCP: http://www.jscp.gr.jp) was established. The JSCP has been working on the dissemination of CP, the standardization of medical care, and the development of electronic CP.

2   Outcome-Oriented Clinical Pathways

At JSCP, the concept of CP was defined as "a method of improving the quality of medical care by analyzing the deviation from the standard which is the standard medical care plan including the patient condition, the goal of medical treatment and evaluation and record". The following description follows this definition.

First, we will explain the main terms (Table 1).

In clinical practice, medical treatment is carried out in response to a variety of patients and clinical processes, but the method is fairly personal. In other words, treatment policies and standardization did not proceed smoothly due to medical experts with disorderly diagnoses, prescriptions, and examinations. Even in the 1990s, there was a lot of preoperative shaving, (presently regarded as malpractice), administration of long-term prophylactic antibiotics, administration of antibiotics after surgery, etc., without concomitant surveys done on the results.

The outcome-oriented CP standardizes different clinical goals, assessment, record system, etc. by medical personnel, and aims to realize high-quality medical treatment. The outcome here is a clinical target. For example, there is an observation of the outcome "pain control is done" (the pain on the first day after surgery is 3 or less on the NRS: Numerical Rating Scale). There is an observation item such as "no fever" in the outcome (body temperature on the third day after operation is 37.0 degrees C or less). If these measured values are [NRS = 6] and [body temperature = 38.2 °C], they are judged to be out of the target value and therefore not appropriate. As such, concurrent pain control can not be done, because of fever and the case becomes a variance. When a variance occurs, the reason should be stated in the medical record and evaluation of why it could not be achieved and a description of countermeasures should be recorded.

Table 1 Terminology in clinical pathway

- Outcome: goal or target in clinical care
- Assessment: evaluation item for an achievement of outcome
- Task: our clinical practice or work to achieve outcome
- OAT unit: Outcome-Assessment-Task as a basic care process
- Variance: a case of no achievement of outcome
- Critical indicator: important outcome to influence the result of treatment
- Overview: one sheet of whole clinical pathway understood by a glance
- Daily based record: 24 h based record with outcome, assessment, task and record including variance

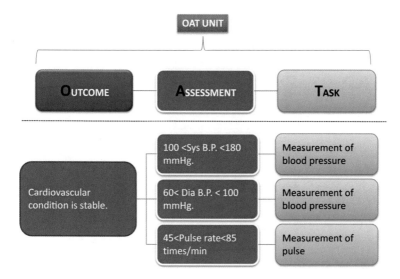

Fig. 1 The basic concept of OAT unit

Assessment (observation items), tasks, and outcomes are a set and form a basic unit of medical care. Here, this basic unit is called an OAT unit. This relationship is illustrated as follows. (Fig. 1).

By establishing this basic unit, the definition of the outcome-assessment (observation item) and task became clear, making it possible to construct a CP data model for computerization.

Currently, the Japan Medical Informatics Association (JAMI: https://www.jami.jp) and the JSCP are jointly building a repository for storing data obtained from CP and medical records. When this is completed, data integration across hospitals will be possible, and even in the case of different EMR vendors, data integrating environments can be constructed, and medical big data can be accumulated. This can be expected to greatly contribute to adoption of CP, creation of new drugs, improvement of medical care, collection of side effect information, and the like.

3 How to Collect Data from CP

The data in CP can be roughly divided into two groups. One is the data of the model OAT Unit before CP implementation, the other is the clinical data obtained after individual CP implementation (Fig. 2).

First of all, I will explain how it is structured. The OAT unit, which is the basic unit of clinical practice, consists of outcome, assessment (observation item), and task (Fig. 3). Combining the basic OAT units as necessary, the overview of CP is completed (Fig. 4). This is the basic illustration of CP, which looks like a framework.

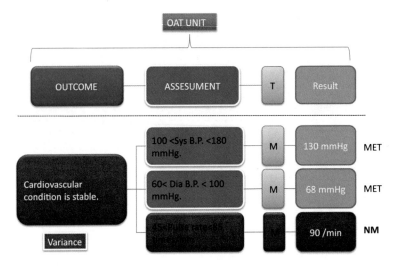

Fig. 2 Model OAT unit and real data

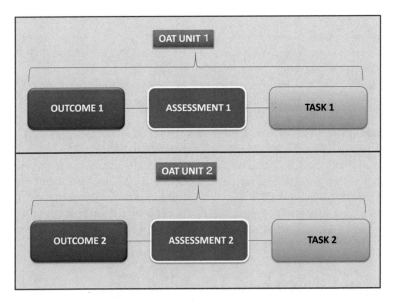

Fig. 3 OAT unit as a basic care process

In accordance with the situation and policies of each hospital in this basic design, adding or deleting outcomes creates CP that seem appropriate for that hospital. Furthermore, when adapting to an actual patient, it is normal to add an outcome according to the patient's condition. No actual data has been collected because CP has yet to reach that stage. In other words, the process of creating a box that stores

Overview

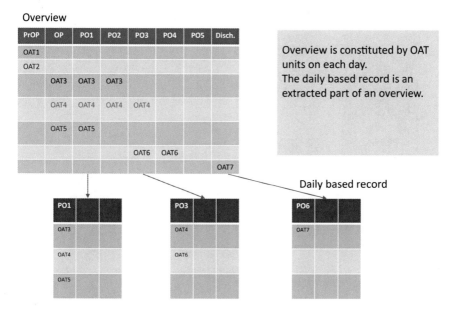

Fig. 4 Overview and daily based record with OAT unit

data in an orderly manner is under way and real data will be written from the clinical site in the future.

After the boxes are formed, real data will be put in, but if you do not manage data input it will not be possible to analyze data. Real data is varied. Although it is relatively easy to process numerical data, textual data is not accurate data which can be scientifically analyzed because it relies on not precisely defined linguistic terms. Therefore, the JSCP standardized outcome terms and observation terms in the OAT unit, and created a master by combining the tasks with MEDIS nursing terms. The BOM (Basic Outcome Master) was created in this manner, so that it became possible to handle data as a structured language, not a natural sentence (Fig. 5).

4 BOM (Basic Outcome Master)

BOM is the master of outcomes upgraded in 2011 by the JSCP. To be precise, "stable circulation" is the superordinate concept of outcome, which has observation items such as blood pressure and pulse, which are subordinate concepts to "stable circulation" (Fig. 6). Usually one outcome has less than three observation items. Most of the clinical conditions are expressed by combining these outcomes with observation items. By using BOM as a standard expression of clinical condition, data collection from records became easier.

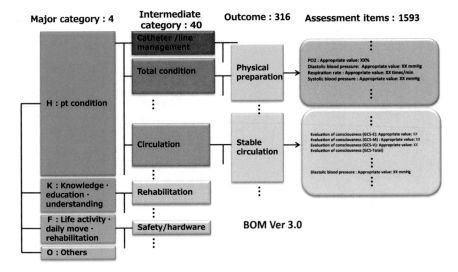

Fig. 5 The basic structure of BOM

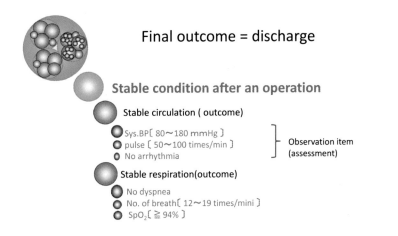

Fig. 6 Outcome and assessment item

The BOM consists of four levels: major classification, middle classification, outcome, and observation items, each consisting of 4, 40, 316, 1593 (Fig. 5). If outcomes and observation items are insufficiently expressed or new treatment emerges and new outcomes and observation items become necessary, they are temporarily classified as "others" and used in the facility. Once a year we collect these "others", review them at academic societies, and add them if necessary. BOM is a standard expression method of patient condition, and as such does not change drastically. Currently, a standardized set of outcome and observation items is being promoted, with outcome "respiration is stable" (no breathing difficulty), (breathing rate is 25 times/minute or less), (no cough and sputum) recommended.

As mentioned above, if even one of the three observation items deviates from the appropriate value, the overall outcome becomes a variance. By using the standard term BOM, we can take patient data accurately and in large quantities. This will accelerate due to digitization and further contribution to clinical research will be possible. Previously it was only possible to review the detailed condition of the patient by rereading charts written in natural language by various types of medical professionals. This complicated work has been eliminated by computerization with BOM.

5   Dynamic Template [2]

The template was developed by Medical Knowledge Sharing (MKS: www.mksinc. co.jp/).

Variance is the most significant factor leading to improvement compared with several other clinical processes. In other words, by analyzing variance, you can understand therapeutic pitfalls and the critical indicators (see Table 1), and also prove the guideline's reliability. Therefore, it is necessary to keep detailed records of variance. In the past, these records were either not taken or taken by each type of medical professional, so it was difficult to gather such information later. Using paper charts, it is difficult to collect vast amounts of information systematically.

In our EMR (MegaOAKHR made by NEC) there is a feature called dynamic template which automatically appears when variance occurs (Fig. 7), and so we will write a record in the SOAP (Subjective, Objective, Assessment and Plan) column. By doing this, it is possible to efficiently record information with 1/10th the conventional numbers of characters. Moreover, since it is possible to record this dynamic template for each SOAP, data collection for variance analysis is easy. Of course, the content written in this column is a natural sentence, and it is also possible to gather this and perform text mining. Furthermore, if you organize the contents of this template partly as a default, it is possible to compile automatically. By analyzing such information, medical support to patient condition change, such as what should be dealt with next variance occurrence, is possible.

6   How to Store the Data

The data thus generated is stored in the DWH, but it can not be analyzed unless the data is stored in a linked form. Especially if events such as hospitalization date, operation day, medication date, examination date, discharge date, etc. are not linked with all data, accurate analysis can not be done efficiently. Variance is especially important and must be recorded. At the moment, such pathway data is basically only collected

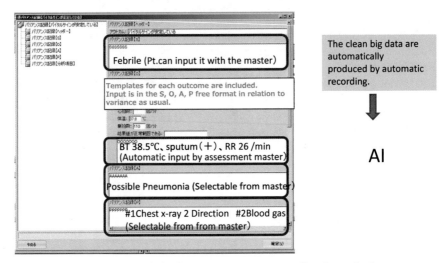

If items of assessment and plan are made into masters, it is easy to collect data and to inspect validity of assessment and treatment.

Fig. 7 Variance recording in the dynamic template

internally in hospitals. One reason is that even if standard outcome terms such as BOM are implemented, the subordinate observation items are slightly different depending on the hospital, department, or doctor. Although variance occurrence is actually measured, its measure is not consistent and therefore, accurate comparison can not be made. In the JSCP standardization committee, basic tools such as the critical indicators for not only the outcome but also the observation items and tasks for judging it are determined. As such, BOM is revised so that comparison can be made on the same scale. When this is completed, judgments of outcome are standardized, so highly accurate variance analysis is possible.

Of course, you can not just upgrade the BOM to gather homogeneous data across hospitals and vendors. If you do not unify the codes of the laboratory tests and medicines currently used in each hospital, extra work such as interposing a conversion code is required, and the cost-effectiveness and precision are also reduced. This is similar to the master of tasks, and the fact that there is no nationally standardized code interferes with accurate data collection.

In the future, in order to analyze pathway data comprehensively and accurately, items necessary for analysis must be enumerated in advance, and it is necessary to prepare a repository including event information, patient attributes and the like. The present idea is to add an interface in each hospital's EMR system and store the data in a hospital repository (Fig. 8).

This data would be stored so that it can be analyzed inside each hospital. Furthermore, this data may be sent to an anonymizing accreditation organization of personal information, anonymously processed, and then analyzed. As of November 2018,

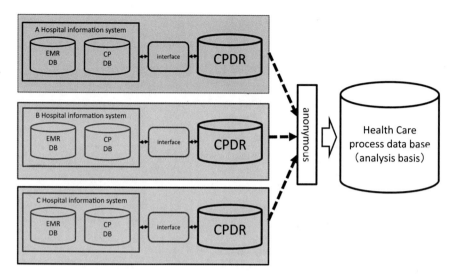

Fig. 8 Interface and standardized data repository of clinical pathway

this accredited institution has not been finalized and it is not certain what kind of mechanism will be used to handle the data.

7 How to Produce Big Data

Big data is data of amounts exceeding the capabilities of ordinary database software, which can be interpreted extensively, both quantitatively and qualitatively. With paper charts or paper based pathways, variance analysis items were limited, and only about 100 cases could be analyzed. For outcome basis BOM data collection, primary analysis is performed for several tens (often about 10–15) of outcomes and about 30–45 observation items accompanying them, and 100–500 target cases data can be easily collected. Furthermore, part of the result is automatically visualized. Of course, not only the observation items but also the medicines used in units of a day and the contents of the examination can be viewed as a table. It seems that this amount of data is probably sufficient to understand medical management. As a result, for example, out of 558 cases of femoral neck fractures, analyses such as these become possible: what is the variance of cases that could not be discharged on schedule? How many cases of fever 37.5 °C or more occurred after surgery including some infection occurred? (Fig. 9). Further, many cases that cannot be discharged on schedule include cases where meals intake was poor and if in these cases pain was above the target value, a program to strengthen pain management can be included on the pathway [3]. It is possible to compare trials for improvements by collecting data and analyzing variance afterwards.

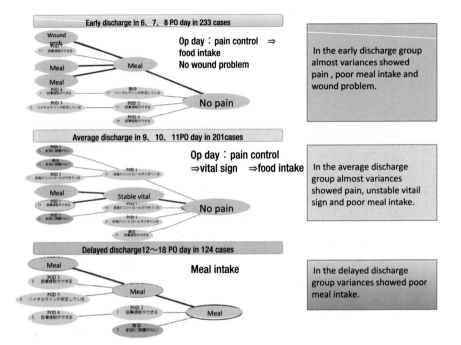

Fig. 9 Text Mining in 558 cases of femoral neck fracture

In order to deepen the primary analysis further upload all the examination values and all the medicines, etc. using the internet and make it even bigger so that we can employ machine learning and make new discoveries.

8 CP and Recording

Dating records in outcome-oriented CP can be said to be presently the most suitable recording system for digitization. In Japan, the conventional paper record is officially known as No. 2 format paper; it is simple to describe symptoms and evaluation in the left column and to describe treatment and prescription in the right column, but there was no further detail describing use. In most of the current EMR systems, laboratory tests or prescription orders are made from each order screen. Laboratory data are confirmed by opening each screen, and prescriptions are confirmed by opening its implementation screen. This format corresponds to an indication sheet or instruction sheet in the paper chart, an examination table, a prescription column, etc. However, it is not consistent with regards to details. Moreover, since accurate times are not stated on paper, there are many ambiguities, such as: records from the previous day, opposite descriptions on the front and back, and lack of clarity as to the writer. Moreover, since it is a natural sentence, in order to extract data from such records, we have to rework everything. Actually, the data that the clinician needs is not just "what", "when", "in what dose", "how long" treatment was administered. More important information includes patient conditions such as "whether pain has been relieved", "whether the circulation dynamics have stabilized", "febrile", "can walk autonomously" and "whether meals were sufficient". When this data is analyzed used a database, we can see trends like "the probability of aspiration pneumonia is higher when there is a fever and cough during hospitalization due to brain stroke" and "if the fever on the 2nd day after admission is above a certain level, the probability of infection is high." (Fig. 10).

In the daily recording sheet the time from 0 o'clock to 24 o'clock, the timing described in the chart, and the date of prescription are all clear. Moreover, the operation day and medications are also clear, as well as the beginning of the discharge process. Because the event information is clear, it is easy to set the starting point of analysis. In other words, all the data described in EMR are associated with the patient state data in chronological order, so that a database is automatically formed (Fig. 11).

Also, other advantages of digitization include automatically triggering examinations, and improvement in the efficiency of recording, such as automatic entry of description field at the time of variance. These data are finalized after expenses are calculated at patient discharge, and can be used as fixed data.

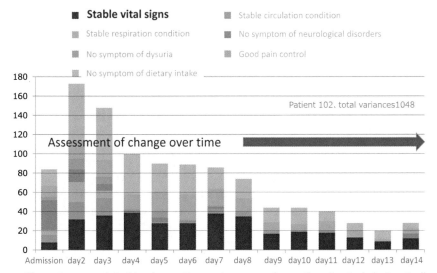

The outcomes related to observation variances can be easily extracted electronically from clinical pathway of brain stroke.

Fig. 10 Brain stroke clinical pathway: variance occurrences over time

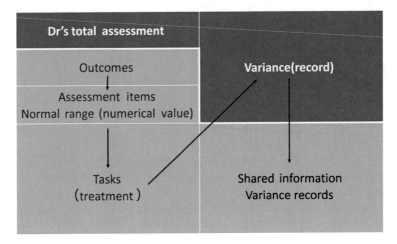

Fig. 11 Structure of daily record

9 How to Analyze the Data and Improve Quality of Medicine

The data created by the electronic pathway is stored in the DWH, and temporary analysis variance distribution, hospitalization number distribution, cost breakdown, etc. are visualized by NECV publicly accessible. Ordinary PDCA can be used for quality improvement at this level, and even improve the efficiency of collecting and analyzing variance. Namely, routine process management is mostly fulfilled at this NECV (Fig. 12).

Of course, if you want perform additional analysis, you can analyze examination, prescription, image report, etc. as big data shared via the internet. For example, to investigate factors influencing discharge 8th day after cerebral infarction in 345 cases, we conducted random forest analysis, one method of machine learning, in collaboration with Kyushu University Medical Information Center [4]. (Fig. 13). When calculating the influence of discharging on the 8th day as the objective variable, and arranging the variables by strength, surprising results are found. If we look deeper, we may be able to abstract factors like those which cause faster recoveries or slow the speed of treatment. This may make it possible to develop new drugs, compare treatments, identify ineffective drugs, and so on.

Although the record at the time of variance occurrence is a natural sentence at the time of recording, since the structure of the template is divided into the columns of S, O, A, P, text mining to obtain conclusions with high accuracy is relatively easy [5].

We have used pathways to periodically collect and analyze variances and improve medical processes. For example, we have tried to improve by choosing the safest treatment method while optimizing for efficacy and cost effectiveness. An example

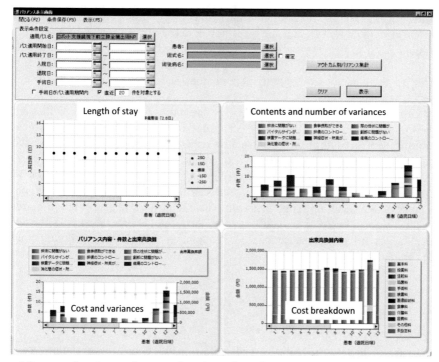

All sorts of variance data can be viewed on the variance display screens at all computer terminals in our hospital.

Fig. 12 NECV (novel electronic clinical pathway viewer)

- ## AUC=0.90

Sufficient forecast performance

Unexpectedly, A/G ratio on the day of admission etc. ranked high.

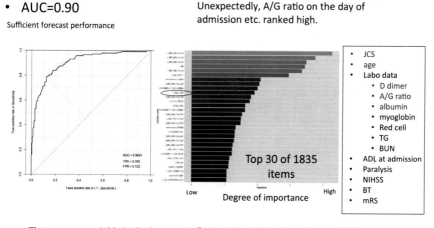

The purpose variable is discharge on 8th day with 1835 explanation variables.

Fig. 13 Example of random forest analysis (cerebral infarction cases)

is the use of antibiotics, analgesics, drain extraction criteria, and postoperative extubation criteria. However, in some cases not all departments show improvement, so through pathway conferences we have worked with our staff and the staff from other institutions to improve overall quality in the region. I have been expanding from learning organizations to learning communities and I am proud with the overall region's progress.

Organizational activities are necessary to promptly perform these activities, and the limited efforts of individuals or departments alone will not lead to significant improvement. In addition, in order to realize improvements, it is necessary to produce high volumes of clean data with the cooperation various individuals, for example, experts in data handling, experts in analysis, and experts in on-site verification and execution. In order to promptly perform activities, efficient data collection, data accuracy, data handling range managed with IT must be implemented in current medical records based on natural language. Therefore, in order for LHS (Learning Health System) to become established, "equipment" and "human resources" must be combined organizationally [6].

10   Future Subject

Activities to improve the quality of medical care include how to prevent human errors, TQM activities, standardization of CP, creation of guidelines, development of new drugs, improvement of nutrition, etc. As a result of effort, treatment outcomes at each stage have improved compared to half a century ago. In the future, big data analysis will support further efforts. Medical records will change from ambiguous descriptions to standardized and structured ones. This will not only increase the accuracy of the recorded data and the accuracy of the description, but it will also create a more readable and accurate record through IoT-supported, automated database creation and generation of natural language text using structured language models such as the BOM. This will be a very "natural" text on the surface similar to that written by medical doctors daily. This will eliminate wasted time by medical professionals creating inaccurate text. It may also solve the issue of labor time in Japanese hospitals, which is quite high compared with other countries. If records are reduced to 1/10th of now, 120 min of nursing charting will be reduced to 12 min, reducing overtime and improving record quality.

As standardization of records further advances, and standard data repositories are developed, data collection beyond the barrier of medical institutions will be possible. In addition, a variety of people will be able to participate in data analysis, which will bring about faster diagnosis and treatment as well as improved quality. If side-effect information can be caught in real time and its solution is presented, medical care in facilities with limited resources will be improved.

It is only recently that large amounts of data can be handled. For this reason, not only the understanding of the general public but also laws regarding information will be inadequate. Personal information is important, but on the other hand there are circumstances where it is hard to handle data if you strengthen security too much. The NDB (National Database) in Japan has been restricted such that the number of cases does not become less than 10 in secondary medical regions and municipalities or 100 cases in cities, towns or villages. It is unknown how well such measures balance personal information protection and data accuracy, and there is no evidence for limits such as 10 or 100. For personal information protection, the important point is that individuals can not be identified, but on the other hand, it seems like the content of the information is more important. Information protection, such as excluding or averaging, is needed for rare diseases, rare congenital diseases, diseases related to privacy such as AIDS, etc. Such data should be made available for processing. Otherwise, it will become difficult to understand these diseases, and treatment and development of new drugs may become difficult.

The basis of accuracy of the database is a unified ID, and since "My Number" is not utilized, database comparison becomes complicated. Another problem is that the data use is limited because access is not widely granted. Such data is also an estimate of sharing, and it seems that a variety of analyses should be carried out from a variety of viewpoints. If we foster accessibility, we will be eclipsed by other countries.

A further major problem is that medical practice masters, such as main examination and prescription masters, are JLAC 10 or JLAC 11, and prescriptions include HOT or YJ codes. Although both are recommended by the Ministry of Health, Labor and Welfare, a variety of codes are used, and problems with data migration and extraction occur even when using the same vendor. If the infrastructure of medical information is not properly maintained, efficient use of data will be hindered. In some sense, it is also a social loss that accumulated data is unavailable or difficult to use. Information infrastructure, such as mandating the use of unified code will be necessary. To the extent that there is an obligation to secure personal information, a universal identification code must also be adopted. If the government is not responsible for maintaining and managing the basic master, a true information society will not flourish.

Acknowledgements I thank Prof. Naoki Nakashima, Dr. Yasunobu Nohara, Dr. Takanori Yamashita, Dr. Yoshifumi Wakata and Dr. Kotarou Mastumoto for their contributions to the design and development of this manuscript. This research has been supported by JAMI, JSCP, and AMED.

References

1. Zander K (1988) Managed care within acute care settings: design and implementation via nursing case management. Health Care Superv 6(2):27–43
2. Matsumura Y (1998) Multi axes data presentation in electronic patient record based on structured data entry. Proceeding of EPRiMP, 188–92
3. Yamashita T, Flanagan B, Wakata Y, Hamai S, Nakashima Y, Iwamoto Y, Nakashima N, Hirokawa S (2015) Visualization of key factor relation in clinical pathway. Procedia Comput Sci 60:342–351

4. Nohara Y, Matsumoto K, Nakashima N (2018) Extracting predictive indicator for prognosis of cerebral infarction using machine learning techniques. Stud Health Technol Inform 245:1280. https://doi.org/10.3233/978-1-61499-830-3-1280

5. Takanori Y, Onimura N, Soejima H, Nakashima N, Hirokawa S (2018) Graph clustering system for text-based records in a clinical pathway. Stud Health Technol Inform 245:649–652. https://doi.org/10.3233/978-1-61499-830-3-649

6. Soejima H, Matsumoto K, Nakashima N, et al (2020) A functional learning health system in Japan: experience with processes and information infrastructure toward continuous health improvement. Learn Health Syst e10252. https://doi.org/10.1002/lrh2.10252

Medical Process Analysis by Using All-Variance Type Outcome-Oriented Electronic Clinical Pathway Data-Exploratory Extracting Critical Indicator

Yoshifumi Wakata

1 Introduction

Outcome-oriented clinical pathway (CP) is a medical process chart which contains favorable patient state to achieve as "outcome" during in-hospital stay. These "outcome" are linked "assessment" to estimate whether "outcome" is achieved or not, objectively and "task" containing medical practice (blood examination, treatment, instruction and so on) to do. Each outcomes and tasks were recorded "variance", if the patient has not achieved favorable state or medical practice has not performed for the patient. Therefore, a sequence of quantitative data concerning medical process from admission to discharge has been easily available by introduction of CP. Therefore, variance data of all patients who was applied CP has been available as structured electronic data. Moreover, electronic CP enable to collect of objective data concerning all patient condition during in-hospital stay and reduce work load to accumulate variance data compared with conventional paper-based CP [1].

2 Critical Indicator

Critical indicator (CI) is one of the CP outcome which causes the crucial influence of the patient clinical outcome such as length of hospital stay (LOS), medical expenditure, destination of discharge and patient satisfaction [2, 3]. Figure 1 indicates that "outcome A" and "outcome C" are CI and "outcome B" is not CI.

Y. Wakata (✉)
Medical Information Management Center, National Hospital Organization Kyushu Medical Center, Fukuoka, Japan
e-mail: wakata@rhythm.ocn.ne.jp

© The Author(s), under exclusive license to Springer Nature Singapore Pte Ltd. 2022   19
N. Nakashima (ed.), *Epidemiologic Research on Real-World Medical Data in Japan*,
SpringerBriefs for Data Scientists and Innovators 2,
https://doi.org/10.1007/978-981-19-1622-9_2

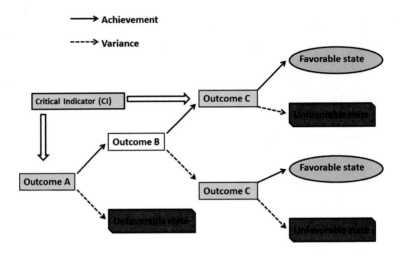

Fig. 1 Critical indicator

Because in spite of how "outcome B" is, when variance occurred at "outcome A" or "outcome C", a patient will be unfavorable result.

3 Analysis of Outcome Oriented CP Data-Exploratory Extracting Critical Indicator (CI)

In the medical process analysis, "What or which process should we put much value on? That is to say, extracting Critical indicator (CI) among varied medical infor-mation" is very important [4]. Importance of CI has been recognized sufficiently. However, setting of CI in producing and using CP are not based on the scientific evidence, because in most cases CI has been empirically chosen from clinical pathway outcome considered important clinically.

Then, we introduced some examples of analytic method using CP data. These methods were constructed in our previous research. First example, we employed traditional statistical analytic method (e.g. general or generalized linear regression model).

Table 1 is the result of the factor correlated LOS of PCI patient. The result shows that age, dietary intake (POD1), abnormality in puncture site (POD1), vital sign stability (POD2) are CIs. In other words, these factors and CP outcomes are risk factors for postponing discharge [5].

Most remarkable advantage of this method is that CIs can be extracted from CP data exhaustively and exploratory compared with the analytic method employed

Table 1 Percutaneous coronary intervention CP analysis result (n = 531)

Variables	Odds ratio	95% CI	P value
Age	1.12	1.02–1.23	0.015
Gender (male = 0, female = 1)	4.39	0.96–19.89	0.060
BMI	1.04	0.89–1.20	0.631
Absence of chest pain (POD1)	5.91	0.41–85.04	0.192
Dietary intake (POD1)	73.52*	1.46–3692.33	0.032
Abnormality in a puncture site (POD1)	23.12*	1.14–467.57	0.041
Vital sign stability (POD2: expected discharge date)	32.55*	1.55–684.94	0.025

Dependent variable: length of stay (within 4 days = 0, over 5 days = 1)
Analytic method: multiple logistic regression

sentinel- or gateway-typed CP. Conventional variance analyses employed sentinel- or gateway-type CP cannot detect all true or potential CI due to limit the CP outcomes variance before analysis. Meanwhile this method enables to extract all true or potential CI among CP outcome. Moreover, when other clinical practice data excepted CP data (e.g. diagnostic-procedure-combination (DPC), SS-MIX2 data which contains drugs and laboratory examination data) will be combined CP data, we may be able to extract CIs from not only on the CP but also outside of the CP. These CIs will be unexpected by healthcare providers (Fig. 2).

One of issue about this method is that this method cannot detect temporal and spatial layered structure of CP outcome. Because our method based on general or generalized linear regression model, it is suitable for estimating independent effect of each factor. But it is not suitable for estimating interaction effect with few or more factors.

Another one is concerned with numbers of explanatory variables. Because this method employs the linear regression model, when the number of explanation variables is enormous, an initial regression model is not able to establish. So we are trying to establish analysis method by using another method such as a machine learning method to solve issue concerning this issue.

Next example, we employed machine learning method. Two analysis samples are introduced in the previous chapter. One is analysis sample of using random forest (Fig. 13 in the previous chapter) [6]. Advantage of this method is that it can put the quite many explanatory variables which are to the extent it can't be handled by a regression model in the analytic model, simultaneously. Additionally, influence of each factor are calculates and ranked by strength of its influence.

Another one is analysis sample of using mind map (Fig. 9 in the previous chapter) [7]. This method can visualize what kind of influence the temporal spatial relation of more than one factor. It is also an advantage of this method that a result is very easy to understand for medical care professionals (doctors, nurses and so on). These two methods can settle the issue of analysis using regression model. In addition, a highly

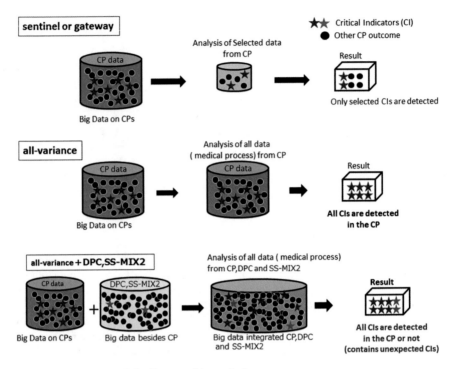

Fig. 2 Data coverage and significance of the analysis

detailed and precise analysis will probably become possible by using the technique of the new mechanical learning, for example deep learning, from now on.

4   Conclusion

Outcome-oriented CP is the only system which can collect high quality structured in hospital patient status data at the current moment. So it can provide good materials of analysis using machine learning and AI.

Moreover, when standardization of outcome oriented CP data has been more developed, more refined visualization and analysis using CP data in many facilities has become possible.

Another important thing is to grasp a relation with medical examination and treatment data besides the CP data sufficiently. When you can integrate those and use it for an analysis, wider knowledge is obtained, those can contribute to medical quality improvement.

Disclosures   There is no COI to declare.

References

1. Wakamiya S, Yamauchi K (2009) What are the standard function of electronic clinical pathways? Int J Med Inform 78(8):543–550
2. Zander K (2001) How clinical pathway positively transform health care organization. Jpn Soc Clin Pathw 3:11–17
3. Whippe TW, Little AB (1997) Variance analysis for care path outcomes management. J Nurs Care Qual 12(1):20–25
4. Weinstein MC, Fineberg HV (1980) Clinical decision analysis. Saunders
5. Wakata Y, Nakashima N (2015) Utilization of clinical pathway data in multi center. In: 35th joint conference on medical informatics. Okinawa. Japan
6. Nohara Y, Matsumoto K, Nakashima N (2017) Extracting predictive indicator for prognosis of cerebral infarction using machine learning techniques. Stud Health Technol Inform 245:1280. https://doi.org/10.3233/978-1-61499-830-3-1280
7. Yamashita T, Flanagan B, Wakata Y, Hamai S, Nakashima Y, Iwamoto Y, Nakashima N, Hirokawa S (2015) Visualization of key factor relation in clinical pathway. Procedia Comput Sci 60:342–351

Information and Data Standard Development for Clinical Pathways

Mihoko Okada, Naoki Nakashima, and Hidehisa Soejima

Keywords Outcome-oriented clinical pathway · Pathway data model · Outcome-assessment-task unit · ePath · Standard ePath message

1 Introduction

Many of the recent hospital information systems (HIS) or Electronic Health Record Systems (EHRS) have certain functionalities to support clinical pathways. But there are no standards of clinical pathway systems or pathway data, which hinders analyses of pathway data or even comparisons of pathways from different clinical institutions. To address the problem, the joint committee of the Japanese Society for Clinical Pathway (JSCP) and the Japan Association for Medical Informatics (JAMI) initiated the electronic outcome-oriented clinical pathway project. "The Development and Use of Standardized Data Model for Clinical Pathways" granted by AMED (Japan Agency for Medical Research and Development) has been undertaken by the joint committee as a two-and-a-half year's term research and development project since October 2018. The goal of the project called "ePath" in short is to establish interoperable electronic clinical pathways that enable outcome analyses and process analyses.

From the view point of information processing, the ePath project aims to develop:

M. Okada (✉)
Institute of Health Data Infrastructure for All, Tokyo, Japan
e-mail: mihoko.okada@idial.or.jp

N. Nakashima
Medical Information Center, Kyushu University Hospital, Fukuoka, Japan

H. Soejima
Saiseikai Kumamoto Hospital, Kumamoto, Japan

© The Author(s), under exclusive license to Springer Nature Singapore Pte Ltd. 2022
N. Nakashima (ed.), *Epidemiologic Research on Real-World Medical Data in Japan*,
SpringerBriefs for Data Scientists and Innovators 2,
https://doi.org/10.1007/978-981-19-1622-9_3

(1)   standard representation of interoperable outcome-oriented electronic pathways (ePath),
(2)   standard pathway information systems that enable transfer of ePath and ePath data, and
(3)   a platform for accumulation and analyses of standardized ePath and related clinical data.

In this chapter, the components of the standards corresponding to the aims (1) to (3) above are described. Hereinafter, "clinical pathway" may be called simply "pathway" in this chapter.

2 Overall System Design of the ePath Project

The OAT (outcome-assessment-task) is a fundamental unit to represent clinical processes. Figure 1 shows a conceptual representation of a sample OAT-unit based pathway.

In the ePath project, OAT-unit based archetypal pathways of eight clinical objectives were developed by the research team composed of researchers of the participating hospitals, co-investigators, and specialized clinicians of supporting academic associations.

Figure 2 shows the overall system design of the ePath project. The four hospitals, Saiseikai-Kumamoto Hospital, Kyushu University Hospital, NTT Medical Center Tokyo, and National Hospital Organization Shikoku Cancer Center have participated, and NEC Corporation, Fujitsu Limited, IBM Japan Ltd, and Software Service, Inc. are the four HIS vendors of the respective hospitals. We describe the main technical components of the ePath project, that is, standard ePath message, ePath data repository, and standard path systems within hospitals for interoperable outcome-oriented electronic pathways (ePath).

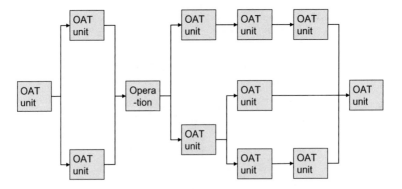

Fig. 1 Conceptual representation of an OAT-unit-based pathway

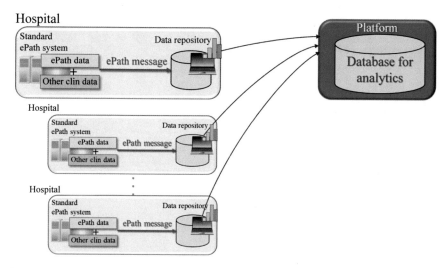

Fig. 2 Overall system design of ePath project

3 ePath Standards

3.1 ePath Message

The standard ePath message is a format to represent, transfer, extract and process ePath data. The ePath message has been developed based on the concept of OAT unit, the combination of "outcome, assessment and task." To give you an idea what the ePath message transfers, Fig. 3 shows an overall outline of the message data model as a class diagram. The class "ePath message" represents hospital information, patient information, admission information and ePath application information. The class "ePath application inf" is composed of applied pathway information and OAT unit. The class "OTA UNIT" is repeatable and is composed of "Outcome", "Assessment", and "Task". It also contains "Overall evaluation". In the class "Outcome", BOM stands for the standard Basic Outcome Master that is developed and maintained by JSCP. Based on data models, ePath message data structure and data elements were designed and the specifications have been developed. The ePath message is implemented using XML currently. Implementation of the ePath message, however, is not limited to XML.

3.2 ePath Data Repository

An ePath data repository has been introduced into each participating hospital for storing ePath data and other clinical data including administrative data, medical

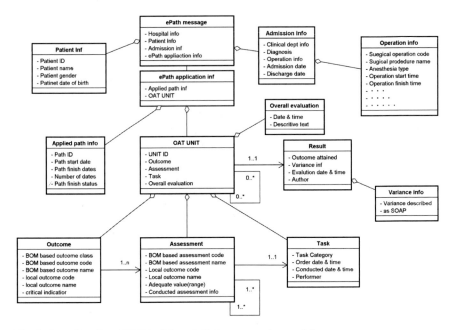

Fig. 3 Overall outline of OAT-unit based ePath message data model

expenses breakdown data, laboratory test data, prescriptions and injections data, and so on. The repository is accompanied by a repository management system that provides fundamental functionalities including user registration, access log recording, ePath data rendering, tabular and graphical summarization of ePath data such as the number of cases for each pathway by gender, by age class, by types of surgical procedures, and so on. An ePath data repository and its management system facilitate management and use of ePath data by the respective hospitals.

3.3 Standard Outcome-Oriented Pathway System

The existing pathway systems of the four participating hospitals have been modified to accommodate OAT-unit based pathways and to generate and transfer ePath message to data repositories. Under IRB approval and authorization of the respective hospitals, the systems have been in operation and OAT-unit based archetypal pathways of eight clinical objectives have been applied to patients with informed consent since April 2019.

In the ePath project, the user interfaces of different pathway systems were also investigated. Standard requirements were defined and the existing systems were modified to meet the requirements. This would allow clinicians who work at more than two institutions or who move to a new institution to use pathway systems in the same or almost the same way regardless of HIS vendors.

3.4 ePath Data Platform

ePath data repository serves two purposes, firstly for ePath data management within a hospital, and in the second for collecting ePath data from multiple hospitals. For the latter purpose, anonymized ePath data and related clinical data have been collected from the repositories of the participating hospitals and stored in the ePath database. More than 4,000 cases are accumulated in the database as of writing. Clinical process data obtained from ePath messages of the four hospitals were compared and discussed by clinicians of the hospitals, and the findings were fed back to the clinical settings. The accumulated data have been and are being processed for visualization and machine learning [1, 2].

4 ePath Project in the Future

Outcome-oriented ePath standards are introduced. To promote interoperable clinical pathways, pathway information systems, pathway data among clinical institutions and HIS vendors, the standard ePath message has been established. Standard ePath message represents, transfers and extracts pathway data of multiple clinical institutions with differing HIS vendors in the standardized data structure and data elements. ePath message enables collection, integration and extensive analyses of accumulated data. The finalized ePath message has been laid out on the table for adoption as a JAMI (Japan Association for Medical Informatics) standard and is expected to be adopted soon. Then it will be applied for adoption as a HELICS (Health Information and Communication Standard Board) standard. With cooperation of JAHIS (Japanese Association of Healthcare Information Systems Industry), ePath message and repository implementation guides are under review.

The electronic OAT-based pathways developed will be uploaded to an electronic ePath library managed by an academic association to allow any hospitals to download and import to their pathway systems. The project will promote interoperable clinical pathways with transportable pathway data among vendors, which will lead to standardization of many other clinical pathways. The ePath based clinical process management is deemed an essential component of the next generation EHR systems.

"The Act on Anonymized Medical Data that are Meant to Contribute to Research and Development in the Medical Field", Next Generation Medical Infrastructure Law in short, was promulgated in May 2017 and became effective in May 2018 in Japan. Under the Law, certified agent for anonymization of patient data is newly introduced. Only agents certified by the competent Minister can be entrusted with non-anonymized patient information, and can collect a large amount of patient data from clinical institutions, anonymize and provide data to users from industry, government or academia. To provide patient information to a certified agent, a clinical institution is required to notify the patients regarding the use of their information and provide a chance to refuse by opt-out. Two agents have been certified as of writing this. The ePath project aims to enable integrated analyses of accumulated pathway data under the Next Generation Medical Infrastructure Law. Data analyses infrastructure constructed upon the law is expected to vitalize clinical pathway activities in small or midsize hospitals and consequently, to improve quality and efficiency of care.

Disclosures   None of the authors have relevant conflicts of interest.

References

1. Soejima H, Matsumoto K, Nakashima N, Nohara Y, Yamashita T, Machida J, Nakaguma H (2020) A functional learning health system in Japan: experience with processes and information infrastructure toward continuous health improvement. LHS J
2. Yamashita T, Wakata Y, Nakaguma H, Nohara Y, Hato S, Kawamura S, Muraoka S, Sugita M, Okada M, Nakashima N, Soejima H (2020) Machine learning for classification of postoperative patient status using standardized medical data. Asia-Pacific association for medical informatics (APAMI2020) proceedings, 131

Standard Code Mapping and Data Quality

Japan Laboratory Code (JLAC) 10

Dongchon Kang

1 History

The Japan Society of Clinical Pathology (predecessor of the current Japan Society of Laboratory Medicine [JSLM]) created the Japan Laboratory Code (JLAC) system mainly for clinical laboratory testing [1]. The coding system was established in 1963, meaning its 60-plus-year history far exceeds that of the Logical Observation Identifiers Names and Codes (LOINC), which started in 1994 [2] and Nomenclature for Properties and Units (NPU) terminology originating in 1995 [3]. The JLAC was originally intended for classification of clinical tests, used within a facility such as a hospital or central clinical laboratory, though not for inter-facility use. JLAC version 10 (hereinafter JLAC10) was formulated in 1997. The latest version, JLAC10.a23 includes minor revisions. In 2011, the Ministry of Health, Labour and Welfare of Japan acknowledged JLAC10 as the Standard Laboratory Test Code Set for clinical information systems.

2 Structure of JLAC10Description of Activity and Work Performed

JLAC10 comprises five elements: analyte code (five alphanumeric characters), identification code (four digits), specimen code (three digits), methodology code (three digits), and result code (two digits) (see Table 1 for an example).

D. Kang (✉)
Department of Clinical Chemistry and Laboratory Medicine, Kyushu University Graduate School, Fukuoka, Japan
e-mail: kang.dongchon.218@m.kyushu-u.ac.jp

© The Author(s), under exclusive license to Springer Nature Singapore Pte Ltd. 2022   33
N. Nakashima (ed.), *Epidemiologic Research on Real-World Medical Data in Japan*,
SpringerBriefs for Data Scientists and Innovators 2,
https://doi.org/10.1007/978-981-19-1622-9_4

Table 1 Example of JLAC10

Serum AST		
JLAC10 code (17 digits)		
3B035000002327301		
Element	Code	Name
Analyte (5)	3B035	AST
Identification (4)	0000	–
Specimen (3)	023	Serum
Method (3)	272	UV absorption
Result (2)	01	Quantity

1. Analyte code. This characterizes types of tests and indicates specific analytes via five alphanumeric characters. As an exception to the standard rule, "substance" may be substituted by "reaction" in some cases; e.g., occult blood reaction, TTT, ZTT.

The characters contain three subsections. The first digit represents the primary classification or type of test, the following letter represents the secondary classification or type of analyte, and the remaining three digits represent the analyte. In Table 1, the analyte code is 3B035; wherein the 3 means "biochemical test," the B means "enzyme," and the final three numerals mean "aspartate aminotransferase, AST."

2. Identification code. When a test needs further classification of the analyte code, a four-digit number is used; e.g., for specification of a virus, antibody, allergen, or cluster of differentiation of a lymphocyte.
3. Specimen code. This uses a three-digit number to show the type of sample used for testing.
4. Methodology code. This uses a three-digit number to show a clear analysis principle or analysis reaction. This code is to ensure comparability of data with the same code.
5. Result code. This explains the properties of the results, though it does not indicate a specific unit such as mg/ml or unit/ml.

The codes for the five elements are independent of one another and predetermined in accordance with bylaws. All codes are listed on the JSLM website (https://jslm.org/committees/code/index.html; in Japanese). Thus, JLAC10 has a logical and flexible structure enabling easy coding and understanding. One clinical test is essentially determined by the analyte and specimen, and then further distinguished by the methodology. If required, the identification gives more detailed classification. Finally, a property of the unit is added using the result. Table 1 shows a simple representative example—the serum AST test. Here, its identification code is "0000" because further classification is not required. Refer to the files (JLAC.zip) written in English in the JSLM website (https://jslm.org/committees/code/index.html; in Japanese) for more details.

3   Operation and Maintenance

As mentioned, JLAC10 was originally intended as a code for use within local facilities for administrative management processes, such as ordering, billing, and recording. To that end, the bylaw describing coding principles is sufficient. Owing to this, in fact, the JLAC10 codes came to vary among facilities (local JLAC10 codes). However, assigning different codes to one test among facilities caused no conflict as long as all coding and usage operations were completed in each facility. The Japanese government currently promotes regional care systems based on cooperation with hospitals and healthcare facilities in the regions. Large-scale medical information databases are also being built nationwide.

Amid this climate, JLAC10 increasingly must serve to provide common codes. Local JLAC10 codes naturally compromise the quality of databases when JLAC10 is used in the databases. A central coding system has been built to remedy this problem (Fig. 1). The Medical Information System Development Center (MEDIS-DC) and Committee on Clinical Laboratory Test Code of JSLM jointly run this JLAC operation center. The center already stocks JLAC10 codes corresponding to

Fig. 1 Japan Laboratory Code (JLAC) operation system

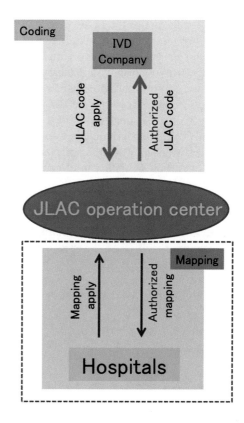

all product documents of approved in vitro diagnostic agents in Japan and promptly provides new product codes.

When an in vitro diagnostic (IVD) company releases a new product, the company asks the JLAC operation center to assign the product a JLAC10 code.

4 JLAC11

As described above, the biggest role of JLAC10 is now as a common coding system for laboratory test data stored in large-scale health databases. From this standpoint, in addition to the central coding system, a central system for mapping (i.e., connection of JLAC10 codes with all laboratory tests conducted in one facility) (Fig. 1, lower part indicated by dashed line) is also needed for standardization of codes used in regional hospitals.

A challenge additional to the standardization is that JLAC10 inherently harbors several deficiencies as a database code because it was originally made for single-facility use. The first problem is that the methodology code does not have sufficient classification granularity to enable precise comparability. This code is an extremely important feature of the JLAC system. Obviously, the measured values in clinical laboratory testing can greatly differ depending on the analytic conditions, particularly for enzymes. The proper methodology code signifies the comparability or commutability of results in databases and therefore, by selecting the same code, collection only of comparable data from among vast amounts of data is easily enabled.

The second issue is that the JLAC10 result code does not indicate units of the results. A single hospital usually has one application for one analyte and therefore can naturally can assign its unit. However, data within a large database derive from many facilities, which may use different types of tests for one analyte. To transform various forms of data into a unified and comparable form, the JLAC code should specify the unit itself (e.g., mg/ml, unit/ml), but not only its properties like quantitative and qualitative.

JLAC10 is currently undergoing a revision to version 11, or JLAC11, to overcome these issues, and resultantly to improve data utility concerning large databases and increase inter-facility usefulness. JLAC11 is aimed at precise use of large databases. The basic structure of JLAC10 will be retained.

Acknowledgements This work is supported in part by a grant of Japan Agency for Medical Research and Development (18mk0101075h).

Disclosures There is no COI to declare.

References

1. Japanese Society of Laboratory Medicine (1969) Committee report: classification code for

central clinical laboratory. Rinsho Byori 17:1–5 (article in Japanese)
2. LOINC. https://loinc.org/about/
3. Rigg JC, Brown SS, Dybker R, Olsen H (1995) Compendium of terminology and nomenclature of properties in clinical laboratory sciences. London, Blackwell Science. https://doi.org/10.1016/S0307-4412(96)80015-X

ICD-10 and ICD-11 in Japan

Takeshi Imai

1 History of ICD and Usage in Japan

The first edition of the International Classification of Diseases (ICD), known as the International List of Cause of Death, was adopted by the International Statistical Institute in 1893, with 179 disease categories for mortality statistics. WHO was then entrusted with the ICD at its creation in 1948, and published the 6th version (ICD-6) that incorporated morbidity for the first time. The ICD has subsequently been revised with more classification of diseases and is now widely used for both mortality and morbidity statistics more than 100 countries around the world. In Japan, the first edition of ICD was applied in 1900, and now the 10th version (ICD-10) has been used since 1990, which includes 14,473 disease categories.

ICD-10 itself is a classification system of disease categories and does not necessarily cover various disease names used in daily clinical practice. In that sense, ICD-10 is not a terminological system. Therefore, in Japan, a standard Japanese disease terminology called ICD-10-based Standard Disease Code Master is widely used in hospital information systems (HIS), not only for diagnosis entries but also for health insurance claims. It includes more than 26,000 standard Japanese disease names with corresponding ICD-10 codes, more than 2,300 modifiers, and more than 100,000 index terms. Of course, not all disease concepts in clinical records are described by such standard disease names. For mortality reporting, Ministry of Health, Labour and Welfare Japan analyzes free-text death certificates using automated coding tools and expert checks. There are also many free text descriptions of diseases, disorders, and

T. Imai (✉)
Center for Disease Biology and Integrative Medicine, Graduate School of Medicine, The University of Tokyo, Tokyo, Japan
e-mail: imai@m.u-tokyo.ac.jp

health conditions in clinical notes and narrative reports. Capturing these phenotypes from clinical data is still an important task for secondary use of clinical real-world database (RWD).

2 ICD-11

In 2007, WHO launched the 11th major revision process with the aim of updating classifications to reflect critical advances in science and medicine and promoting an electronic use, and in June 2018, the 11th version (ICD-11) was officially released. It contains around 17,000 diagnostic categories and more than 100,000 medical diagnostic index terms, a significant increase over ICD-10. ICD-11 is fully electronic and a suite of web services improves ease and accuracy of coding requiring less user training than ever before. ICD-11 is significantly different from the previous ones and includes various features as shown below.

(A)   Description of the properties of each disease category

Until ICD-10, ICD was just a classification system of diseases, however in ICD-11, semantic properties of each disease category are described according to the Content Model that includes anatomy, manifestations, causes, temporal and severity properties, functions, and so on. As of 2018, only limited items of the Content Model are described, however, the description of these properties is planned to be enhanced in future version of ICD11. It is expected to be used as a knowledge base on the clinical characteristics of various diseases, and to enhance cooperation with information described in electronic medical records.

(B)   Coding mechanism

The codes of the ICD-11 are alphanumeric and covers the range 1A00.00 to ZZ9Z.ZZ, which is a significant change from the ICD-10 notation. In ICD-11, the structure of the classification was also significantly changed from ICD-10 and it allows "multiple parenting" to improve usability. For example, gestational diabetes belongs to multiple upper categories such as "endocrine, nutritional or metabolic diseases" and "pregnancy, childbirth or the puerperium".

ICD-11 not only provides a classification system for diseases, but also include, as much as possible, fined-grained terms and synonyms that belong to each disease category. Each of those pre-coordinated terms corresponds to one stem code. In addition, when it is desired to describe more detailed health condition that cannot be coded by a pre-coordinated term, it can be represented by combining two or more stem codes or combining stem codes with one or more extension codes as modifiers. This mechanism is called Post-Coordination. For example, the code for "diabetic retinopathy due to Type2 diabetes mellitus" can be represented as "9B71.0Z/5A11" by combining the code 9B71.0Z (diabetic retinopathy) and 5A11 (Type2 diabetes mellitus). In the same way, lung cancer has the code 2C25.Z, while right lung cancer can be represented as "2C25.Z & XK9K" by combining the extension code XK9K for "right". ICD-11 includes Chapter X as a set of 14,000 extension codes in various categories such as severity, etiology, anatomical details, and histopathology, and so on. Using post-coordination mechanism, it is possible to code a detailed health condition at any level.

(C)   New chapters

ICD-11 also includes new chapters other than Chapter X (extension codes) explained above. For example, supplementary chapter for traditional medicine conditions was added as Chapter 26. The chapter refers to disorders and patterns which originated in ancient Chinese Medicine and commonly used in China, Japan, Korea and elsewhere around the world, but not intended for mortality reporting.

Another important supplementary chapter is Chapter V. For the purpose of functioning assessment, International Classification of Functioning, Disability and Health (ICF) exists, but has not been fully utilized. This chapter was newly included in ICD-11 for the purpose of linking with ICF, and it may assist to identify severity of diseases and disorders, and to describe the impact of the health condition on the daily life of a person as a functioning set. The set of functioning items in Chapter V allows the WHO Disability Assessment Scale (WHODAS), and the Model Disability Survey (MDS) to be used to generate a summary functioning score. This is expected to facilitate the recording of functioning assessments of daily life, for care needs assessment and outcome evaluation of interventions.

3   Application of ICD-11 to Japan and Future Prospects

ICD-11 will enable us to code patients' conditions with a finer granularity than ever before using post-coordination mechanism. In addition, with the introduction of Chapter V, the foundation for facilitating data collection on functioning assessments of daily life is also being prepared. Proper use of ICD-11 in the near future is thought to play an important role in the formation of high-quality healthcare big data.

On the other hand, considering the burden of performing detailed coding in clinical settings and living environments, it will be necessary to devise ways to collect high-quality, detailed, and wide-ranging clinical data without increasing the workload as much as possible. It is also considered necessary to radically review the current electronic health record (EHR) system and the way of clinical information recording. Currently, ICD11 is in the process of being translated into Japanese in cooperation with the Ministry of Health, Labour and Welfare and academic societies, and its application to Japan still needs to take a few years. However, towards the application of ICD-11 to Japan in the near future, it is thought that it is time to reconsider creating a mechanism to generate high-quality medical big data by utilizing its potential as much as possible.

Standard Codes for Prescribing Drugs Use Multiple Code Systems Depending on Their Purpose

Atsushi Takada

In Standardized Structured Medical Information eXchange 2 (SS-MIX2), the standard contract in Electronic Data Capture (EDC) of Japan, HOT code is specified as the standard code of drug [1]. The HOT code is a code system that contains all 13 digits, comprising 7 digits (A) + 2 digits (B) + 2 digits (C) + 2 digits (D), based on the purpose of a drug. The first seven digits (A) signify the particle size to recommend the medicine; i.e., it expresses the ingredients and standards of the drug. This is called "HOT 7," and it can be used for collecting medicines sold by multiple companies as a group. The next two digits (B) specify the sales company. Along with the top seven digits, it is named "HOT 9" as nine digits, and the medicine itself can be itemized. HOT 9 has the same granularity as the YJ code (described later) and is used for various purposes, including prescribing drug prices. The next two digits (C) identify the package form. This is called "HOT 11." In Japan, it is feasible to differentiate the mainstream Press Through Pack (PTP) sheets, packaging types, such as bottle packing, and the number of contents. The next two digits (D) imply the version, corresponding to GS 1-Databar code, which is a distribution code of a medicine. Overall, this is referred to as "HOT 13," which corresponds to the change in the distribution code, enabling to specify each drug to be sold. Besides the HOT code, YJ code, a drug price list code (commonly known as the Ministry of Health, Labour, and Welfare code), a drug classification code, GS 1-Databar code, and the like are used. The drug price list code is determined by the Ministry of Health, Labour and Welfare, which is a regulatory agency for pharmaceuticals, and is used to ascertain the drug price, which is the official price.

The code system has seven numerical values plus alphabetic characters plus four numerical values.

A. Takada (✉)
Medical Information Center, Kyushu University, Fukuoka, Japan
e-mail: takada.atsushi.721@m.kyushu-u.ac.jp

© The Author(s), under exclusive license to Springer Nature Singapore Pte Ltd. 2022   43
N. Nakashima (ed.), *Epidemiologic Research on Real-World Medical Data in Japan*,
SpringerBriefs for Data Scientists and Innovators 2,
https://doi.org/10.1007/978-981-19-1622-9_6

In Japan, two types of drug price listing exist, which is listed by brand name and unified name.

In the brand name listing, as the drug price is determined for each medicine, this code can specify the medicine. In the unified name listing, however, drugs cannot be identified because the same drug prices are prescribed for the same drugs. Hence, the YJ code is expanded so that medicines can be recognized by expanding the drug price standard listed code [2]. The medicinal-substance classification code constitutes the first four digits of the drug price–based receipt code, and it outlines a medicinal-effect group. The GS1-Databar code is used in circulation, which is stipulated in the packaging unit. The code has a branch number in each dispensing unit, packaging unit, packing unit, and is used in various situations.

- When to change the codes

  The standard codes of these drugs are modified because of various factors. Owing to its characteristics, the distribution code is changed when packaging is changed. Otherwise, it is not feasible to differentiate between old packaging and new packaging. Similarly, although a distribution code exists for each packaging unit, it is not a suitable code to specify a prescription.

  The code associated with the drug price is changed not by changing the drug price itself but when changing the drug name. Besides, it will be changed even when the sales company of the drug is changed. In that case, it will be changed as follows. First, a part can recognize the sales company in the distribution code, following which the distribution code is changed. Then, as several medicines include a character string for determining a sales company called a "shop name" in the product name, the code associated with the drug price is changed. In many cases, this is not executed at the same time but is done sequentially.

- Follow the changes

  In the medical information system, it is essential to follow changes in these drug codes. Although these changes should be made in real time as much as possible, it is essential to consider the distribution stock inside hospitals, and the correspondence is considered to be different for each hospital [3].

Disclosures There is no COI to declare. There was no financial support from any external organization. This article was written by the author alone.

References

1. HELICS. HEaLth Information and Communication Standards Board. http://helics.umin.ac.jp/files/HS001/HS001_rep20180525_MEDIS.pdf
2. Iyaku Joho Kenkyujo. https://www.iyaku.info/yjcode/ (in Japanese)
3. Japan Association for Medical Informatics Healthcare Information Technologist Certification (2016) Shinoharashinsya Publishers. Iryoujyouhou 5th edn: the book of medical information system (in Japanese)

Data Quality Governance Experience at the MID-NET Project

Jinsang Park

1 Development of Procedure for Unified Management of Standard Codes

Recently, with the widespread implementation of "Real-World Data (RWD)," there is increasing interest in using high-quality administrative claims, laboratory tests, electronic health records (EHRs), and information-sharing resources to enhance detection of serious adverse drug reactions (ADRs) [1]. For example, in May 2008, the US Food and Drug Administration launched the Sentinel Initiative, a multi-year program for the establishment of a national electronic monitoring system for medical product safety [2]. In Japan, the Medical Information Database Network Project (MID-NET) has been launched to establish a nationwide medical information database and to advance safe systems for pharmaceuticals using active surveillance of pharmacoepidemiological data in FY2018 [3, 4].

The MID-NET project (previously known as the "Japanese Sentinel Project") aims to promote effective safety measures for drugs by scientifically taking safety measures in pharmacoepidemiological methods utilizing RWD. The MID-NET project database included information of RWD collected approximately 5.05 million patients (as of March 2020) from 10 medical institutions, including 23 hospitals in Japan. Also, the Standardized Structured Medical Record Information eXchange Version2 (SS-MIX2)-based standardized database system was based on EHRs of hospitals for analysis and evaluation of ADR [5]. One of the most important conditions for conducting pharmacoepidemiological research using medical information among medical institutions is to systematically evaluate the quality of the data provided. However, the MID-NET Validation Project found that prioritizing the

J. Park (✉)
Department of Pharmaceutical Sciences, School of Pharmacy at Fukuoka, International University of Health and Welfare, Fukuoka, Japan
e-mail: park21@iuhw.ac.jp

© The Author(s), under exclusive license to Springer Nature Singapore Pte Ltd. 2022   45
N. Nakashima (ed.), *Epidemiologic Research on Real-World Medical Data in Japan*,
SpringerBriefs for Data Scientists and Innovators 2,
https://doi.org/10.1007/978-981-19-1622-9_7

management of the medical institution's local codes about medical services, which resulted in delays in standard codes mapping or omissions in standardized data [6]. Additionally, the Hospital Information System (HIS) depends on a specific EHR vendor and SS-MIX2 organization. Based on these findings, to precisely continue data quality management of using new standard codes in databases of multiple medical institutions and standard codes management, our research group received support from the Japan Agency for Medical Research and Development (AMED) to establish the governance center at the Kyushu University Hospital that could collect and manage information about the standard code interoperability.

2 Effect of Data Quality Management Through a Governance Center

The MID-NET-integrated data sources include 11 types of currently available standard codes, such as the International Classification of Diseases 10th Revision (ICD10) [7], Japan Pharmaceutical Standard Codes (called the HOT, YJ) [8, 9], and Japan Laboratory Test Standard Code 10th Revision (called the JLAC10) [10]. These standard codes are based on SS-MIX2 standard storage. The governance center collects all 11 types of standard codes data to be analyzed from the medical data of multiple medical institutions and constructs a centralized database that mainly controls data quality. The table used when saving a MID-NET-integrated data source is referred to as a "mapping table." The consistency of the standard codes was evaluated using a mapping table from each participating medical institution from the database of the governance center. Then, consistency is monitored, and observations were fed back to each cooperating medical institution (Fig. 1).

We requested the provision of a mapping table for cooperating medical institutions at the national level (23 hospitals of 10 medical institutions) to investigate the standard code mapping ratios. We defined standardization of codes based on a mapping table used in the MID-NET project, and a mapping table of each medical institution was collected by the Governance Center. Then, we examined the mapping ratio status of the JLAC10 code at each medical institution. According to the aggregate results of the JLAC10 code mapping ratio in MID-NET of the cooperating medical institutions, >2,000 JLAC10 codes were mapped to the three cooperating institutions (including hospital groups), but in several institutions, the total number of JLAC10 codes mapped was <100. Moreover, there was a discrepancy among the JLAC10 codes. The reasons for inconsistencies in JLAC10 in the mapping table at each institution can be categorized as follows: (1) low JLAC10 code numbering ratio in the mapping table, (2) difference between JLAC10 operations among cooperating institutions, (3) laboratory code revision history not managed (equipment change, outsourcer change, etc.), and (4) consistency between the mapping tables was not reflected in real-time.

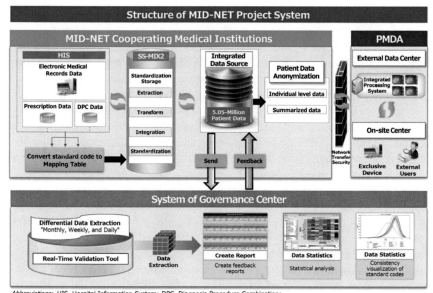

Fig. 1 The framework of governance center and simulation of real-time validation tool

As shown above, many medical institutions prioritize local code management for medical practice, causing real-time delays and omissions in the mapping work with standard codes and maintenance management. Moreover, we found that the operations and HISs of each medical institution significantly affect the mapping of the standard codes. Our first approach to the above problems was to use real-time verification information for standard code mapping and data quality management.

3   Development of Real-Time Validation Tool for Central Governance

As described above, by focusing on mapping activities of standard codes, to precisely continue data quality management in real-time and evaluate new standard codes in multiple medical databases and plan standard codes management, we developed a "real-time validation tool" to manage standard codes and introduced it into three MID-NET cooperating medical institutions. The framework of our simulation of this real-time validation tool is shown in Fig. 1. The mechanism of the real-time validation tool is to automatically output differential information (standard code change, addition, etc.) "monthly, weekly, or daily" in 11 kinds of standard codes by comparing the past-extracted code that was backed up from the HIS with the newly extracted code. Using the real-time validation tool, we were able to visualize situations where differences in standard code mapping occurred essentially.

Degradation of data quality of the standard code, as this factor is temporary, continuously, and suddenly occurs at each institution; thus, integrated management is extremely difficult, resulting in poor data quality. Additionally, standard code items that are not listed in the mapping table should be coded and mapped by a central governance organization. This central governance approach shows that consistency with the use of the standard code by the governance center effectively improved the data quality management system by detecting and unifying the mapping situation of the standard code in the MID-NET project. Taken together, our findings reinforce the importance of real-time management for standardized data and show that data quality management can be improved by integrating and managing standard codes using a central governance method. While data inconsistencies were found in the initial stage, data quality was dramatically improved through collaborative efforts between cooperating medical institutions and the Pharmaceuticals and Medical Devices Agency. Furthermore, central governance management is expected to effectively improve the standardization of data in epidemiological research projects using real-world data in the EHR system of various medical institutions, such as the MID-NET project.

Disclosures   There is no conflict of interest to declare.

References

1. Jensen PB, Jensen LJ, Brunak S (2012) Mining electronic health records: towards better research applications and clinical care. Nat Rev Genet 13(6):395–405
2. Behrman RE, Benner JS, Brown JS, McClellan M, Woodcock J, Platt R (2011) Developing the sentinel system—a national resource for evidence development. N Engl J Med 458 364(6):498–499
3. Yamada K, Itoh M, Fujimura Y, Kimura M, Murata K, Nakashima N, Nakayama M, Ohe K, Orii T, Sueoka E, Suzuki T, Yokoi H, Ishiguro C, Uyama Y (2019) MID-NET® project group. The utilization and challenges of Japan's MID-NET® medical information database network in post marketing drug safety assessments: a summary of pilot pharmacoepidemiological Studies. Pharmacoepidemiol Drug Saf 28(5):601–608
4. Yamaguchi M, Inomata S, Harada S, Matsuzaki Y, Kawaguchi M, Ujibe M, Kishiba M, Fujimura Y, Kimura M, Murata K, Nakashima N, Nakayama M, Ohe K, Orii T, Sueoka E, Suzuki T, Yokoi H, Takahashi F, Uyama Y (2019) Establishment of the MID-NET® medical information database network as a reliable and valuable database for drug safety assessments in Japan. Pharmacoepidemiol Drug Saf 28(10):1395–1404
5. Matoba T, Kohro T, Fujita H, Nakayama M, Kiyosue A, Miyamoto Y, Nishimura K, Hashimoto H, Antoku Y, Nakashima N, Ohe K, Ogawa H, Tsutsui H, Nagai R (2019) Architecture of the Japan ischemic heart disease multimodal prospective data acquisition for precision treatment (J-IMPACT) system. Int Heart J 60(2):264–270
6. Nakashima N (2016) Pharmaceutical regulatory harmonization and evaluation research project "Practical analysis method and education on drug epidemiology research for benefits and risk assessment of pharmaceuticals using MID-NET". The Japan agency for medical research and development (AMED) outsourced research and development results report [Grant Number: 15mk0101014h0102] [Japanese]

7. The World Health Organization (WHO) (2017) ICD-10 version (classification of diseases)
8. Kaihara S, Takekuma R, Tsuchiya F (2002) Standard master for pharmaceutical products (HOT reference number). JAMI 22(4):315–319
9. Ministry of Health, Labour and Welfare (2019) List of drug prices and information on generic drugs [Japanese] [internet] [cited Nov 09, 2019]. Available from: https://www.mhlw.go.jp/topics/2019/08/tp20190819-01.html
10. The Japan Society of Laboratory Medicine (2019) The Japan laboratory accreditation cooperation (JLAC) 10 [internet] [cited Feb 18, 2019]. Available from: https://www.jslm.org/committees/code/index.html

Phenotyping

Phenotyping in Japan

Tatsuo Hiramatsu

When using data from medical records, disease conditions and patient information that are not clearly described must be determined. For example, even if the clinical information for the medical treatment fee is accurately described in the administrative claims data, it does not always precisely indicate the medical condition of a patient. Unlike clinical registries, medical records often do not contain the information you want to obtain. In addition, you may not collect the information that you need from the clinical registry. Sometimes you must stop obtaining information due to the limited number of acquirable items. To cope with the unavailability of the needed data, information must be identified and estimated using plurality of other items in the data, which is referred to as phenotyping.

Phenotyping was first developed in the United States in the late 2000s as part of genomic medicine research. Once a blood sample is acquired, genetic information can be systematically and automatically obtained one after the other, and the medical personnel assessing the status of each patient can register clinical information one by one to the repository. If this gap continues, an overwhelming difference will occur in the speed of storing genetic and clinical information, which will be a restriction to the development of genomic medicine research. To quickly obtain a large amount of clinical information and to link such information to genetic information to conduct statistical analysis for the identification of relationships, efforts began with the idea that clinical condition might be automatically identified with a computer based on the existing electronic medical records even if it is somewhat less accurate than manual assessment. Phenotyping itself is not based on genomic medicine methods. However, it is derived from the field of genomic medicine. Thus, the term phenotype is similar to the genomic field.

T. Hiramatsu (✉)
Department of Medical Informatics, International University of Health and Welfare, Tokyo, Japan
e-mail: hiram-1@umin.net

© The Author(s), under exclusive license to Springer Nature Singapore Pte Ltd. 2022   53
N. Nakashima (ed.), *Epidemiologic Research on Real-World Medical Data in Japan*,
SpringerBriefs for Data Scientists and Innovators 2,
https://doi.org/10.1007/978-981-19-1622-9_8

1 Real-World Data, Designed Data, and Role of Phenotyping

Data used in phenotyping are often real-world data, which are generated during medical care. Such data include those from electronic medical records and administrative claims data. They are not collected with the intention of using in later analysis. By contrast, in inferential statistics where the relationship between items is analyzed, analysis results can be interpreted based on the assumption that the data are obtained as designed in accordance with the usage intention. It should be recognized that such kind of data, designed data, and real-world data are different and that such data have varying natures, even if the appearance is represented in the same form.

Phenotyping converts real-world data into designed data. This action will allow for the interpretation of the results of inferential statistics. In that sense, whether all real-world data items, including those that are not processed with phenotyping, should be considered as designed data. This examination is called validation. The validation of the items obtained using phenotyping is discussed in the later part of this study (Fig. 1).

A slightly different expression is also possible. In the first place, values in designed data are obtained by measuring real-world itself according to usage intention. Different usage intentions lead to different measurement methods, which is the reason why a prior design is necessary. For example, in terms of physical measurement, whether it is gross or net depends on the intention usage of the measured value. A more obvious example would be measurement using questionnaire. Because the words in the question measure what is in the mind of the respondent, the method of measurement (that is, the wording of the question) is cautiously adjusted according to the intention of what you want to obtain.

Values in real-world data are records of facts that occurred according to the way they are generated, and they have no prior design for secondary use. Thus, posterior adjustments or confirmations must be made, and phenotyping should be performed as necessary.

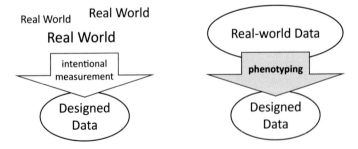

Fig. 1 Phenotyping and designed data

2 Three Stages that Put the Use of Clinical Data on Track

The preparation for the secondary use of accumulated actual medical care data can be divided into three stages of data flow. The first stage involves the time of medical information generation and primary use. At stage 2, the recorded data are organized for secondary use and stored in a storage or database. Assessment of data quality and standardization of data format and representation will be performed at this stage. Stage 3 involves appropriate interpretation of organized data that will be utilized for analysis by understanding the origin of each data and interpreting the expressed semantic range. Phenotyping is one of the activities in stage 3.

It would be good if the information necessary for secondary use in stage 1 can be recorded without losing it. However, in actual medical care and hospital work, the labor and cost become an issue, and it cannot be easily identified. Aside from time and effort, the necessary information cannot be comprehensively recorded in advance, without knowing what kind of data will be used later. Therefore, information is not always recorded at stage 1.

While the purpose of utilizing results after such stages varies, the stages themselves are common. This can be compared to the relationship between a satellite and launching of a space rocket. Similar to satellites with various purposes, such as those for weather, communication, and space exploration, that can be placed on the target orbit by a common multistage rocket, the use of clinical data is on track with the three course stages (Fig. 2).

3 Methods of Phenotyping

Phenotyping methods are categorized, and the specific categories are introduced below. A combination of these categories may be used. In the following description, the judgment is described on the premise that it is binary. However, it may be multi-value or continuous numerical value.

(1) Conditional expression

To determine whether a target disease state is present, a plurality of matching conditions is combined. For example, when there is "HbA1c value of 6.5% or more" or "use of a medicine for treating diabetes," it can be used as a method of determining that the disease is diabetes. It is a traditional method that has been used even before the development of phenotyping. However, in phenotyping, there are numerous cases in which source data are time-lapse data (such as those from electronic medical records). Thus, a judgment must be made with conditions that include a time axis, such as date of clinical laboratory test or duration of diabetes prevalence.

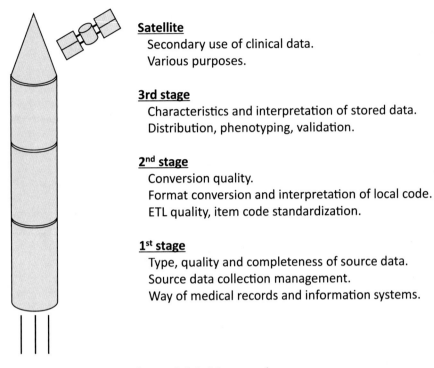

Satellite
  Secondary use of clinical data.
  Various purposes.

3rd stage
  Characteristics and interpretation of stored data.
  Distribution, phenotyping, validation.

2nd stage
  Conversion quality.
  Format conversion and interpretation of local code.
  ETL quality, item code standardization.

1st stage
  Type, quality and completeness of source data.
  Source data collection management.
  Way of medical records and information systems.

Fig. 2 Three stages that put the use of clinical data on track

(2) Algorithmic/rule base

Providing step-by-step instructions to identify the status of a patient is a method of determination according to an algorithm. The procedure is often represented by a flowchart, which can be easily programmed using a computer. Thus, this method is a good match with phenotyping, and the original phenotyping assumes this method. More complex rules compared with conditional expressions can be developed. Unlike machine learning, rule base determination is also excellent in terms of stability and reliability against new data when rules can be made close to the definition of a disease.

(3) Autonomic/machine learning and AI

Instead of a human individually describing the determination method, the computer generates the determination method. In practice, a judgment model is prepared by a human, and the parameters in the model are automatically calculated. For example, in the case of logistic discrimination, a logistic model formula is prepared by a human, and the computer automatically calculates the coefficient of each term. In addition, in deep learning, which has been rapidly developed in recent years, humans develop the structure of the neural network, and the parameters are automatically learned.

Logistic discrimination has been performed for a long time. When machine learning methods, such as SVM and random forest, are used, quantitative interpretation of the model parameters becomes challenging. However, discrimination is often performed more accurately than logistic discrimination. With deep learning, good discrimination results have been obtained particularly in those dealing with image data.

In either case, supervised learning is almost always the case. In supervised learning, a human must prepare training data (which are accurately determined), and preparing a large number of training data often becomes an issue. To prevent this task, the approach of unsupervised learning may occasionally be taken when possible depending on the research theme.

4 Evaluation of Phenotyping

Regardless of which method is used, the accuracy of phenotyping results must be evaluated. The obtained evaluation may be used to search for an accurate phenotyping method compared with another method. It can also be used as reference for the interpretation of analysis results using phenotyping.

A typical method for evaluation is as follows: the phenotyping results are randomly sampled and compared with the correct answer, and indicators, such as sensitivity, specificity, and positive predictive value (PPV), are calculated to show performance. To obtain the correct answer, a medical chart review is conducted. In the review, human experts check the patient's chart one by one and assess the correct answer, and this referred to as the golden standard.

Several points must be considered, which are as follows: how many patient data are appropriate for sampling? What are the proper indicators for calculation? Is one expert or several experts needed? Is there any chance that the judgment wavers among the experts? Do you make a criterion for chart review judgment? How can the criterion be made? In the first place, who are qualified as chart review experts?

The number of sampling must be set so that at least 100 cases are included. When phenotyping rare patient conditions, several review targets are required, which makes operation challenging. Even in such case, when you only need PPV, reviewing appropriate patients via phenotyping will work. Indicators must depend on the purpose of using the phenotyping results.

The criteria for chart review should be set in advance. There is an issue whether to use criteria based on the presence or absence of an objectively clear description or read and judge the implicit context of a medical record. The former is good in that it can make clear decisions. However, questions may arise whether a patient who is supposed to be using that criterion can be pick up. That is, picking up patients based on that criterion might not make sense in the clinical point of view. In the latter, there is a problem as the criteria are dependent on individual expertise and the reproducibility is low. However, one may think that the reality of medical care is that the judgment is somewhat different depending on the expert, and coping with that situation is the way of handling real-world data. Let us think in terms of statistical error. The criteria based on the presence or absence of a clear description are methods with small dispersion and concern about systematic bias. The method of reading the context is opposite, and the dispersion is large. However, the center of gravity is considered as the true judgment if there are several judges to one patient. In reality, there are few phenotyping projects that include a large number of expert judges for chart review of all sampled patient data. However, an alternative plan that included preparing a criteria using clear description and letting several experts with small number of chart reviews evaluate such criteria might be possible.

There is no standard about the kind of expert to judge. Chart reviews are carried out in each project by various professionals, such as specialist physician of a disease, physicians not in charge in the field, specialist nurses, and trained verification experts who are not medical practitioners (including individuals with source document verification [SDV] work in clinical trial). If no objective criteria are set and the context is read by chart reviewers, the judgments of specialist physicians of a disease will be highly preferred.

The most significant disadvantage of using the chart review method is the amount of effort. If not impossible, it is hardly possible to let experts review thousands of medical records for most phenotyping projects. Therefore, a correct answer is sought using other methods other than through chart review. For example, if there is a disease registry for the subject's disease and some patients are registered in such registry, the disease registry can be used to obtain the correct answer. However, for this purpose, linkage between disease registry and phenotyping project data may be possible. In numerous cases, not only technical but also policy issues arise from linkage with other databases.

5   What is the Best Phenotyping Algorithm?

In general, in machine learning, various estimation models are often compared and evaluated to search for a better estimation model. Is it the same in phenotyping? Of course the models and algorithm are important. However, when evaluating phenotyping, we will use some experimental data sets and validate the results against the set that is considered correct. That is, the obtained accuracy is only valid with the data set and the correct set used in the evaluation. Phenotyping algorithm built with the data obtained from one hospital does not usually work well with data from other hospitals. In general, phenotyping algorithm studies about data from multiple healthcare facilities must be conducted to avoid bias due to specific facilities.

This may be considered as a generalizability issue. However, it is important to note that what is true in the first place depends on the purpose. Even if there is an extremely good algorithm with the national data that should be the best in terms of generalization possibility, it does not always have good accuracy in your hospital. This is an important issue if you are conducting phenotyping in a project to improve the quality of medical care at your hospital.

Similar problems occur not only in spatial extent but also in temporal extent. Visiting the healthcare facilities of patients are influenced by medical information in real life. Thus, the characteristics and backgrounds of patients gradually change as years pass by. It is particularly likely to occur if there is a change in the health care system. In addition, the correct answer might be updated with the progress in medicine. Even the diagnostic criteria advocated by the academic society may change. A flexible response will be needed for these issues generated by changes occurring in real-world.

6 Outcome Validation in Pharmacoepidemiology and Phenotyping

Similar to phenotyping, there is outcome validation in the field of pharmacoepidemiology. That is, the use of a drug of interest is considered exposure, and the resulting adverse event is the outcome. The background for outcome validation in pharmacoepidemiology is an area where clinical trials are conducted. Clinical trials are required to ensure the credibility of everything covered in the trial to guarantee the outcome. Similarly, even when using databases instead of conducting clinical trials, the analysis results are used for drug regulation, and it is natural to secure the reliability of data. In such fields, the evaluation of reliability and validity is also essential for the method of extracting outcomes, such as adverse events from medical records. In the field of clinical trials, the term computerized system validation (CSV) confirms if the computer system used is properly developed, installed, and operated. Outcome validation is different from CSV, and it evaluates the validity of classification data, not system, of the patient derived by a certain definition, such as a conditional expression.

Adverse events treated in the field of pharmacoepidemiology are also one of the patient's conditions, and they are essentially the same as the names of diseases mainly dealt with in phenotyping. The name of the disease found in medical records is not necessarily reliable. Furthermore, as the nature of the medical records, the presence or absence of the disease name or the patient's condition that will be focused later for analysis, such as mild adverse events, is not explicitly described. Thus, adverse event extraction and phenotyping share the same idea, which is to solve the problem about the lack of explicit description about the desired information using the other information in the data set.

By contrast, the result of phenotyping can be exposure factor or target population criterion in general clinical research utilizing database as well as outcome factor, which is the use of phenotyping in genomic medicine research. For example, when analyzing the change in the risk of myocardial infarction using suffering from asthma in individuals with type 2 diabetes, phenotyping is necessary in all subjects, exposure, and outcome.

Disclosures There is no COI to declare.

Phenotyping of Administrative Claims Data

Tatsuo Hiramatsu

Administrative claims data (ACD) in Japan is a set of medical treatment information in a healthcare facility and is prepared for medical expenses claim to public health insurance. ACD is created every month at every hospital and clinic that practice insurance medical care. As for clinical laboratory examination, examinations were performed. However, data about the examination results are not available due to claim purposes.

In Japan, all the people are subscribed to one of multiple public health insurances. Furthermore, with any health insurance providers, the insurance coverage criteria are similar, and the claims data are stored in a common format across the country regardless of insurance organization. This situation is extremely different from the situation of electronic medical records from hospital information systems. Therefore, analysis of the ACD may target health treatment of the whole nation. In the past, ACD was usually submitted in paper. However, as a result of requiring electronic form and actively working on digitization, almost all ACDs have been digitized. In this study, we assumed that ACD is in electronic form.

The ACD received by all insurers from medical facilities is submitted to the Ministry of Health, Labour and Welfare (MHLW) under the law regarding health policy making. With the use of these data, the MHLW has built a large anonymous database, which can be used for research after an examination has been conducted by the review committee. In addition, there are also other ACD databases from multiple data source, such as the local government or insurers established by large companies for their employees. It is also possible to individually collect anonymous data from each healthcare facility and to build database of anonymous ACD in accordance with the personal information protection law.

T. Hiramatsu (✉)
Department of Medical Informatics, International University of Health and Welfare, Tokyo, Japan
e-mail: hiram-1@umin.net

© The Author(s), under exclusive license to Springer Nature Singapore Pte Ltd. 2022   61
N. Nakashima (ed.), *Epidemiologic Research on Real-World Medical Data in Japan*,
SpringerBriefs for Data Scientists and Innovators 2,
https://doi.org/10.1007/978-981-19-1622-9_9

One of the problems in using ACD for medical analysis is the reliability of the name of the disease. As it is created for insurance, all the names necessary for insurance claims are listed. However, they do not necessarily represent the patient's medical condition accurately. Conversely, unnecessary disease names for insurance are often not listed in ACD. Because description is considered biased to a specific direction, its use for analysis is inappropriate. Therefore, phenotyping must be conducted to determine whether each patient has the disease of interest.

The advantages of phenotyping from ACD are as follows: data format and data granularity are determined and uniform throughout the country. In the case of ACD from insurers, the database includes all information from all medical facilities that were visited by the subscriber.

The disadvantage is that the data granularity may sometimes be rougher than desired. Because they are grouped by billing unit, multiple medical care activities may be grouped into one billing item. The form and items of the ACD may be updated by the year. Thus, it is necessary to track how the change has been made. In the case of ACD from healthcare facilities, information about the care practiced at other facilities is not available. People usually visit different healthcare facilities based on diseases during the same period.

The evaluation of the result of phenotyping from ACD can be devised in various ways. As each healthcare facility has result values from clinical laboratory examination that is not listed in ACD, result values may also be available from other data source. In such case, result values can be used to obtain a more correct answer. It may not be called golden standard but rather silver standard, which can also be used as training data for machine learning phenotyping and as a correct answer in validation. In particular, it is useful for the type of diseases that can be roughly diagnosed using the clinical laboratory examination result values. Similarly, when Diagnosis Procedure Combination (DPC) data created at large hospitals are available, verification may be carried out by making an item, such as cancer stage, in the DPC data as the correct answer. If cancer registries and other disease registries are available at some healthcare facilities, they may be validated by creating data from the registries as the correct answer. When these ideas are feasible, a lot of efforts can be saved for the evaluation of phenotyping results compared to chart review.

As with phenotyping from ACD or when analyzing ACD, the claims data file is not in an easy-to-analyze format, which cannot be handled by the statistical software. Thus, it is necessary to considerably devise and process the file. One solution is to use a tool that converts ACD files into a general purpose clinical information storage format. There is a format called OMOP Common Data Model, which is used for storing clinical information. This format was originally developed by a joint partnership of the Food and Drug Administration, multiple pharmaceutical companies, and healthcare providers in the United States. However, it is currently maintained and managed by the OHDSI Community, which is an international network of researchers and observational health databases with a central coordinating center at Columbia University. The OMOP Common Data Model can be used by anyone for free. An OMOP conversion tool has been developed for Japanese data, including ACD. Thus, please search for OHDSI Japan. The abilities of Japanese are important because the tool handles ACD written in Japanese.

Disclosures There is no COI to declare.

A Phenotyping Study Using MID-NET Database

Rieko Izukura

1 Introduction

In recent years, clinical research has shifted from interventional/experimental studies to an observational approach using vast amounts of medical data, which is expected to contribute to health promotion or disease prevention. However, such data are often insufficient for observational studies, particularly for describing "phenotypes" (i.e., an organism's observable physical properties or characteristics), because the data were originally produced and accumulated for routine work, such as medical service fee claims. Thus, methods for detecting populations with the targeted phenotypes, i.e., novel phenotyping methods, should be developed to promote the quality of data-driven studies [1]. In Japan, MID-NET, a huge domestic database, is used to monitor the safety of various medical products such as drugs, vaccines, and medical devices [2]. The development of phenotyping methods using data available on MIID-NET is essential to enhance the quality of medical product safety surveillance. In this work, we first describe a trial phenotyping method developed using MID-NET and then outline the future prospects of phenotyping studies using this database.

2 Background: Phenotyping Method Development Using MID-NET

Discussion of phenotyping methods using various medical data, including electronic medical records and domestic administrative claims, was initiated by the Pharmaceuticals and Medical Devices Agency (PMDA) in 2010 [3, 4] and has continued

R. Izukura (✉)
Medical Information Center, Kyushu University Hospital, Fukuoka, Japan
e-mail: izukura.rieko.250@m.kyushu-u.ac.jp

© The Author(s), under exclusive license to Springer Nature Singapore Pte Ltd. 2022   65
N. Nakashima (ed.), *Epidemiologic Research on Real-World Medical Data in Japan*,
SpringerBriefs for Data Scientists and Innovators 2,
https://doi.org/10.1007/978-981-19-1622-9_10

since the operation of MID-NET officially began in April 2018. PMDA has currently developed several phenotyping algorithms in collaboration with medical institutions participating in the MID-NET project. These algorithms can be viewed on the PMDA website, although access permission is required [5].

3 Trial: Development of Phenotyping Algorithms to Identify Interstitial Pneumonia

Interstitial pneumonia (IP) is considered a drug-induced adverse event. The most effective method of diagnosing IP is chest X-ray or computed tomography (CT) scan. To identify as many IP patients as possible in the MID-NET dataset, we needed a phenotyping method to identify patients who were not registered as having IP, although IP was indicated on their CT scans. Thus, we validated a phenotyping method to identify both possible and potential IP groups using MID-NET data.

Data source

MID-NET provided a sampling frame of 117,401 patients hospitalized at Kyushu University Hospital, Fukuoka, Japan between January 1, 2014, and December 31, 2015.

Phenotyping method

The phenotyping method is described below. The left side describes the developed phenotyping method, while the right side describes its application to IP.

Development of phenotyping methods	Application of the developed phenotyping algorithm to IP
Step 1: Creating the main phenotyping algorithm	**Step 1**: Creating the main IP phenotyping algorithm
1. Creation of initial rule-based phenotyping algorithm A using medical guidelines	1. Initial extraction of algorithm A: (IP-related ICD10 codes, e.g., J841) or (KL-6 $\geq$ 430 U/mL)
2. Data extraction from MID-NET and random sampling	2. Of 117,401 cases, extraction of 1,424 cases and random sampling of 200 cases
3. Medical chart review by a medical expert and positive predictive value (PPV) calculation to determine validity	3. Two experts' independent chart reviews of the 200 cases (weighted κ coefficient = 0.74). True positive cases (TP): $n = 73$ (PPV = 36.5%)
4. Estimation of a predictive model for IP with the highest area under the curve (AUC) using machine learning techniques	4. Estimation of a predictive model for IP using the gradient boosting decision tree (GBDT) and r-part function. (Objective variables: TP = 73, explanatory variables: 1,406 structured EMRs, AUC = 0.798)

(continued)

(continued)

Development of phenotyping methods	Application of the developed phenotyping algorithm to IP
5. Modification of initial phenotyping algorithm A: Modified algorithm **A′**	5. Modified algorithm A′: (initial algorithm A) and (KL-6 ≥ 145 U/mL) or (IP-related DPC diagnostic code) (PPV = 80.0%, sensitivity within 200 cases = 60.3%)
Step 2: Creating a sub-phenotyping algorithm	**Step 2**: Creating a sub-phenotyping algorithm
6. Extraction of further possible cases not identified by the initial algorithm A using structured or nonstructured EMRs	6. Extraction of 3,963 possible IP cases without initial extraction algorithm Use of eight text words (e.g., 間質性, すりガラス in Japanese) in the CT reports and then random sampling of 84 cases
7. Medical chart review by a medical expert and PPV calculation	7. Two experts' independent chart reviews of the 84 cases (weighted κ coefficient = 0.61). TP cases: $n = 39$ (PPV = 36.5%)
8. Creation of a sub-phenotyping algorithm **B** using machine learning techniques	8. We did not create a sub-extraction algorithm using GBDT owing to low AUC (i.e., 0.582)
Final phenotyping algorithm: Modified algorithm **A′** OR sub-phenotyping	**Final IP phenotyping algorithm**: The same as [modified algorithm A′ (STEP1-5)]

Although a sub-IP algorithm was not created in this study, we may eventually find unknown IP cases not detected using the initial algorithm A. Further validation of phenotyping methods for detecting potential IP cases is thus required.

4 Future Prospects of Phenotyping Studies Using MID-NET

With the development of artificial intelligence (AI), clinical studies using electronic medical data will increase. Similar to the US Food and Drug Administration, drug safety assessments using developed phenotypic algorithms will enable data-driven research based on scientific validity and reliability. Eventually, this approach will contribute to improving the quality of safety surveillance.

Acknowledgements This study was supported by Japan Agency for Medical Research and Development (AMED) Grant Number 17mk0101088h0001: Research on data characterization and outcome validation for promoting phamacoepidemiological study utilizing MID-NET® for benefit-risk assessments. This contents was presented at the 22nd Japan Association for Medical Informatics conference in 2018 [6].

Disclosures There are no conflicts of interest to declare.

Ethic consideration This trial was approved by the Ethics Review Committee of Kyushu University (29-167).

References

1. Kimura E (2018) Considerations of the observational research database utilizing real world data. J Natl Inst Public Health 67(2):179–190
2. Pharmaceuticals and Medical Devices Agency (2018). Pharmaceuticals and Medical Devices Safety Information No. 351. http://www.pmda.go.jp/files/000223348.pdf#page=4. Accessed 23 Nov 2018
3. Pharmaceuticals and Medical Devices Agency. MIHARI Archive (In Japanese). http://www.pmda.go.jp/safety/surveillance-analysis/0007.html. Accessed 23 Nov 2018
4. Ishiguro C, Takeuchi Y, Yamada K, Komamine M, Uyama Y (2015) The progress of MIHARI: medical information for risk assessment initiative project in PMDA and new paradigm of pharmaco-vigilance in Japan. Jpn J Pharmacoepidemiol 20(1):3–13
5. Pharmaceuticals and Medical Devices Agency. MID-NET (Medical Information Database Network) (In Japanese). https://www.pmda.go.jp/safety/mid-net/0001.html. Accessed 23 Nov 2018
6. Izukura R, Nohara Y, Yamashita T, Hamada H, Suzuki K, Fukuyama S, Matsumoto K, Park J, Takada A, Wakata Y, Kandabashi T, Nakanishi Y, Uyama Y, Nakashima N (2018) Establishment of phenotyping to detect the diseases using the structured data on hospital information system. In: Poster presented at: 22nd Japan association for medical informatics, 21–23 June 2018, Nigata

Integration of Phenotyping Algorithms in Japan

Rieko Izukura

1 Introduction

The development of algorithms for identifying targeted phenotypes is critical in data-driven studies and various collection system of phenotyping algorithm is gradually being constructed worldwide. In the United States, the Electronic Medical Records and Genomics Network has compiled developed algorithms as a database and has released it for public use on the Web [1]. Conversely, in Japan, although the formulation of phenotyping methods and algorithms using domestic database has been attempted in a few academic fields such as pharmacoepidemiology, these have been not integrated to date. In this chapter, we attempt to collect and formatively organize developed phenotyping algorithms and outline this study and future prospects for the integration of phenotyping algorithms.

2 Short Report: Phenotyping Algorithm Collection in AMED Projects

This is one AMED project, "The study about the dissemination of data standardization for benefit-risk assessment of medicines using MID-NET (representative: Naoki Nakashima Kyushu University)".

R. Izukura (✉)
Medical Information Center, Kyushu University Hospital, Fukuoka, Japan
e-mail: izukura.rieko.250@m.kyushu-u.ac.jp

2.1   Study Methods

The phenotyping algorithms were collected following the questionnaires (Table 1) from research resources (Tables 2). The research resources included the annual

Table 1 Questionnaires for phenotyping algorithm collection

Questionnaire
Selected phenotyping algorithm development would include as the follow:
• The purpose of phenotyping algorithm development
• Doctors' chart reviews for identification of true disease diagnostic cases
• Positive predictive value or sensitivity as algorithm validity (If not, specifying "NO CALCULATION" in the list)

Table 2   Research resources

Category	Resources	Criteria	n^*
Project reports	Two projects of Japan Agency for Medical Research and Development (AMED)	Manual search	34
	Medical Information Database Network (MID-NET)		
	Medical Information for Risk Assessment Initiative (MIHARI)		
Literature	PubMed	(Phenotyping OR valid*) AND (positive predictive value OR sensitivity) AND database (Japanese Society for Pharmacoepidemiology http://www.jspe.jp/committee/020/0271_1/) Exclusion of development of phenotyping algorithms without using Japanese databases from extracted literatures	8
	Igaku Chuo Zasshi*		
	*Under the management of NPO Japan Medical Abstracts Society		
		Manual search <exclusion> systematic review	
Academic conference papers	Japan Association for Medical Informatics	Keywords: "data validation," "phenotyping," and "phenotyping algorithm" in Japanese	2

* n, number of phenotyping algorithms

study reports of research projects and the academic conference papers because of inadequate numbers of phenotyping algorithm developed in Japan.

2.2 Results

Table 2 summarizes the results between 2015 and 2018. Forty-four phenotyping algorithms were extracted, including 34 algorithms that have not been published. There were 10 algorithms from literature and conference papers [2–5].

2.2.1 ICD Classification

Among all ICD categories, the phenotyping algorithm of "Endocrine, nutritional and metabolic diseases" (20.5%) was the largest. Conversely, the algorithms of "Psychiatric disease" and "Dermatologic disease" have not been found as yet. These diseases are primarily diagnosed by patient's symptoms or medical images or its reports. In current data-driven studies, it is extremely difficult to identify diseases using these non-structured EMRs. Thus, development of phenotyping algorithm combining natural language processing methods has gradually been performed in Japan (Fig. 1 and Table 3).

2.2.2 Disease Diagnosis

The phenotyping algorithm of "Gastrointestinal hemorrhage" or "Diabetes mellitus" definition showed the largest numbers of ICD disease diagnosis ($n = 4$), followed by "Acute myocardial infarction" ($n = 3$).

3 Future Prospects for Integration of Phenotyping Algorithms

Academic society-driven dissemination activities have been gradually conducted in Japan due to an insufficient experience with development of phenotyping algorithms. The annual symposium held by the Japan Association for Medical Informatics introduced an outline for phenotyping study and its methods. Similarly, the Japanese Society for Pharmacoepidemiology compiled the methodology or issues related to data validation studies using a domestic database and published it on the Web [2]. By further development of phenotyping algorithms and integration of them, easy detection of cohorts (groups) with certain diseases will be feasible using various domestic databases, thus contributing to further development of epidemiological studies.

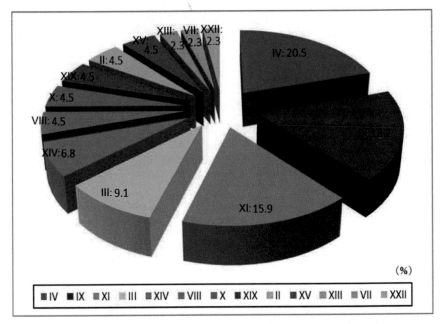

II: Neoplasms, III: Diseases of the blood and blood-forming organs and certain disorders involving the immune mechanism, IV: Endocrine, nutritional, and metabolic diseases, VII : Diseases of the eye and adnexa, VIII: Diseases of the ear and mastoid process, IX: Diseases of the circulatory system, X: Diseases of the respiratory system, XI: Diseases of the digestive system, XIII: Diseases of the musculoskeletal system and connective tissue, XIV: Diseases of the genitourinary system, XV: Pregnancy, childbirth, and the puerperium, XXII: Codes for special purposes, XIX: Injury, poisoning, and certain other consequences of external causes

Fig. 1 Number of phenotyping algorithms for each ICD10 category

Table 3 Calculation of phenotyping algorithms for each disease

Disease diagnosis	N	Disease diagnosis	N
Gastrointestinal hemorrhage	4	Bleeding required blood transfusion	1
Diabetes mellitus	4	Hypoacusis	1
Acute myocardial infarction	3	Intraocular hemorrhage	1
Breast cancer	2	Hypertension	1
Agranulocytosis	2	Intracranial infarction	1
Hyperglycemia	2	Venous thromboembolism	1
Intracranial hemorrhage	2	Perforation of the digestive tract	1
Interstitial pneumonia	2	Acute pancreatitis	1
Acute renal failure	2	Drug-induced liver dysfunction	1
Pregnancy	2	Rhabdomyolysis	1
Anaphylaxis	2	Drug-induced renal disturbance	1
Disseminated intravascular coagulation	1	Hyperlipidemia	1
Thrombopenia	1	Hypocalcemia	1
Hyperthyroidism	1	Auditory impairment	1

Acknowledgements This study was supported by Japan Agency for Medical Research and Development (AMED) Grant Number 18mk0101064h0003: The Study about the dissemination of data standardization for benefit-risk assessment of medicines using MID-NET.

Disclosures There is no COI to declare.

References

1. Phenotype Knowledge Base (PheKB). https://www.phekb.org/(2018)
2. Sato I, Yagata H, Ohashi Y (2015) The accuracy of Japanese claims data in identifying breast cancer cases. Biol Pharm Bull, 38(1):53–57
3. Kagawa R, Kawazoe Y, Shinohara E, Imai T, Oe K (2016) The development and comparison with phenotyping algorithms of hypertension and other disease. Jpn Assoc Med. Informatics, 36(2):770–773
4. Hanatani T, Sai K, Tohkin M, Segawa K, Kimura M, Hori K, Kawakami J, Saito Y (2013) An algorithm for the identification of heparin-induced thrombocytopenia using a medical information database. J Clin Pharm Ther, 38(5):423–428
5. Yamaguchi T, Fuji T, Akagi M, Abe Y, Nakamura M, Yamada N, Oda E, Matsubayashi D, Ota K, Kobayashi M, Matsui D, Kaburagi J, Matsushita Y, Harada A (2015) The Epidemiological Study of Venous Thromboembolism and Bleeding Events Using a Japanese Healthcare Database–Validation Study—, Iyakuhin Johogaku,17(2):87–93
6. Iwagami M, Aoki K, Akazawa G, Ishiguro E, Imai S, Ooba N, Kusaba M, Koide D, Goto A, Kobayashi N, Sato I, Nakane S, Miyazaki M, Kubota K (2018) Nihon ni okeru shōbyō-mei o chūshin to suru reseputo jōhō kara e rareru shihyō no baridēshon ni kansuru hōkoku-sho (Reports: Task force for validation studies about clinical index obtained by medical claims data including disease diagnosis name in Japan). Japanese Society for Pharmacoepidemiology. http://www.jspe.jp/committee/020/0271_1/. Accessed 23 Nov 2018

Data Analysis on Real World Data

Analysis on Real-World Data: An Overview

Tomohiro Shinozaki and Yutaka Matsuyama

1 What Can We Do with Real-World Data?

1.1 Real-World Medical Databases Compared to Conventional Study Designs

The mainstream of clinical research designs has been the *randomized controlled trial* (RCT). In an RCT, two or more distinct interventions are randomly assigned to participating patients who have met strict eligibility criteria, with the patients' outcomes followed-up and compared between the intervention groups. Since such an intentional assignment of interventions is not always feasible from ethical or logistical viewpoints, epidemiologists have been developing the theory and principles of observational research designs including the *cohort study* and many types of *case-control study* designs [1]. The development of medical informatics has brought a new approach to medical research, namely the analysis of massive databases of electronic health records or claims data that include real-world data (RWD) of clinical practice.

The RWD databases often include large (e.g., >100,000) numbers of patients, long follow-up durations, and frequent measurements of clinical data, which can be more cost-efficient in data collection than RCTs and epidemiological designs. As another feature, RWD reflects treatment policies in clinical practice. Hence, RWD analysis could evaluate the *effectiveness* of treatments or medical interventions that would be obtained in real clinical settings, whereas experimental designs like RCTs typically target *efficacy* that would be observed under ideal situations where all patients

T. Shinozaki (✉)
Department of Information and Computer Technology, Faculty of Engineering, Tokyo University of Science, Tokyo, Japan
e-mail: shinozaki@rs.tus.ac.jp

Y. Matsuyama
Department of Biostatistics, School of Public Health, The University of Tokyo, Tokyo, Japan

adhere to assigned treatments (possibly including placebo) throughout the observation period [2]. Although "pragmatic" RCTs or epidemiological cohort/case-control studies sometimes incorporate a broad range of participants, they are markedly more expensive than databases and must set eligibility criteria to achieve specific study objectives.

By definition, the analysis of RWD is essentially within an observational-study framework like cohort and case-control studies. Thus, it is helpful to follow principles and practice in analysis of such epidemiologic data [1]. However, unlike pre-planned cohort/case-control studies, RWD are not collected for a specific study purpose. Hence, we should be more cautious with data quality in the analysis stage. In the following section, we highlight four practical concerns regarding bias in statistical inference from RWD. Before proceeding to these specific statistical topics, we must recognize two distinct purposes of RWD analysis: prediction and causal inference.

1.2 Distinguish the Purposes: Prediction Versus Causal Inference

One of the overarching goals of clinical studies is to establish effective treatment strategies for individual patients. To achieve this goal, researchers may need to determine patients' prognosis from their characteristics, medical conditions, and/or treatment history, and then decide on the best interventional treatment among the available options based on their prognostic information. *Prediction* is the statistical methodology that addresses prognosis and *causal inference* is the methodology that addresses treatment/intervention [3]. Although RWD analyses for both purposes share statistical techniques, it is necessary to distinguish them to select an appropriate analysis strategy.

Prediction is a traditional application and many good statistics textbooks cover the methods for developing and evaluating statistical models for prediction (e.g., [4–6]). The objective is good prediction in future patients, leaving their treatment practice aside; the research questions include "What will happen with this patient in real clinical settings if we treat her/him as usual?" Typically, prediction-model development employs regression models (e.g., linear, logistic, and Cox models) with likelihood-based estimation, but sparse modeling methods (such as lasso and ridge regression) [7] and other machine learning techniques (e.g., random forest, neural network, and support vector machine) [6] would be useful and powerful alternatives with modern computing circumstances. After developing prediction models, their evaluation from several viewpoints is an important next step. For event or time-to-event clinical outcomes, such viewpoints (and corresponding metrics) may include discrimination (sensitivity, specificity, concordance index), calibration (calibration-in-the-large, calibration slope, calibration plot), reclassification (net reclassification and integrated discrimination improvements), and decision analytic utility (net benefit, decision curve) [5, 8].

The question like "Which would be better for this patient if treatment A had been continued or if treatment A had been discontinued?" is counter to the predictive research questions mentioned earlier. This hypothetical *counterfactual* argument formally defines causal effects of treatment A within contemporary theories called statistical causal inference [9–11]. Note that counterfactual comparison of the same patients under distinct conditions (e.g., with and without treatment) is different from comparing observed distinct groups of patients (e.g., treated and untreated groups). That discrepancy is called *confounding*, which is extensively discussed in the following sections. Causal inference methodology provides statistical methods to connect counterfactually defined effects with observed data distributions, revealing the sets of assumptions that would suffice to draw valid conclusions. The aforementioned efficacy and effectiveness are at both ends of the "effects" of treatment policies, depending on how they are determined based on each patient's clinical process during treatments [9, 12, 13].

In the next section, we introduce four bias sources in RWD. While the first three (missing data, measurement error, and nonrandom selection) are related to both prediction and causal inference analyses, the last (confounding) is the concept for causal inference.

2   Sources of and Measures Against Biases in RWD

2.1   Missing Data

Although the volume and variety of RWD databases are often large, the fact that databases are not initiated to address specific research objectives can exclude necessary information that is needed in the analysis. In both prediction and causal inference analyses, all or part of the variables may be unmeasured; they are termed missing variables.

Although we cannot directly use completely unmeasured variables for primary outcome or treatment variables, one may use a *proxy* of them. For example, a treatment exclusively indicated for a specific disease may become the sign of diagnosis of that disease in claims databases. However, as proxy-variables are not a direct measure of the variables of interest, they generally suffer from measurement error [3], which is covered in the next subsection.

Partly unobserved variables can be imputed in some sense by statistical models [14]. Some authors recommend *multiple imputation* for missing data directly, which stochastically imputes data according to estimated statistical models, repeats imputation and creates multiple datasets, and summarizes estimates from the multiple imputed datasets. Another approach is *weighting* complete-case observations whose variables are completely measured in a database, via the inverse probability of measuring all variables predicted by, for instance, logistic models. While it is possible to combine imputation and weighting methods in actual analysis, the statistical principle behind that combination is uncertain.

Note that outcome *censoring* should be distinguished from missing, which is addressed by survival analysis methods (e.g., [15]). Unless censoring is independent of patients' future prognosis such that censored and uncensored patients have the same outcome rate, however, survival methods would fail to converge to unbiased prediction or effect parameters. In the presence of censoring dependent on (possibly time-dependent) measured risk factors, the inverse probability of censoring weights estimated by the risk factors will diminish the bias [16, 17]. Analysts may have to consider *competing risks*, which prevent the occurrence of all or part of other events of interest, by using the notion of subdistribution hazards and their modeling. However, caution may be needed with causal analysis [18].

2.2 Measurement Error

Almost all clinical variables include measurement error irrespective of study designs. For RWD, the proxy-variables described earlier may particularly suffer from measurement error or misclassification relative to the variables of interest. For example, the diagnosis procedure combination (DPC) database includes patients' clinical information and actual treatment procedures; however, it is possible that part of the diagnosis/procedure data are misclassified to obtain more reimbursement, because one of the primary purpose of DPC is to implement an electronic invoicing system for payment [19, 20].

Treatment variables may also suffer from severe measurement error in RWD analysis. While one of the advantages of RWD is information on a treatment process adopted in real clinical settings (which enables the estimation of "effectiveness"-type treatment effects if adequate statistical adjustments are employed), such a treatment process usually changes over time. If patients are allocated to groups according to their treatment status at baseline, their treatment information can be misclassified during the follow-up period [3]. To circumvent the bias due to this type of treatment misclassification, time-varying information on treatment should be recorded and analyzed by landmark analysis or time-dependent survival models (for prediction), or by causal inference methods for sustained/repeated treatments, as discussed subsequently. Confounder information is also likely to be mismeasured or substituted by proxy variables due to missing data, especially for time-varying treatment–outcome relationships. Such measurement errors in adjustment variables result in residual confounding [3, 21].

2.3 Nonrandom Sample Selection

RWD databases typically cover nationwide and broad patient populations. It is widely believed that such *representativeness* of the data assures *generalizability* of study results to a source population. However, equating the representativeness of data to the wide applicability of the results is dangerous [22], especially for RWD analysis, because the temptation for broad sample inclusion with mild eligibility criteria using RWD often conflicts with a rigid analysis conduct. For example, if we face the missing or severely mismeasured variables for some research hypotheses, we may have to change the research question and impose strict criteria for patients to analyze [13].

We underscore the *transportability* of the results, which is intended to predict the results in external target populations (the "as-is" manner in prediction and hypothetical counterfactual manner in causal inference frameworks). We cannot obtain transportable predictions through only representativeness for a "whole" population, such as all Japanese patients at some time points. The target populations will change over time and the results will only be applicable to patients within a restricted time and area. Unless a study sample is randomly selected from a target population every time, stratified analysis by outcome-predictors (in prediction) or effect-modifiers (in causal inference), between which the treatment effects differ, is warranted for broadly applicable results in terms of transportability [22]. As explained in the next subsection, stratified analysis is also a key strategy for addressing confounding in a causal inference framework. Some authors call the difference in outcome-predictor distributions between reference and target populations as also "confounding" that makes the results non-transportable in prediction framework [3].

Finally, despite prospectively collected information in RWD databases, "retrospective" cohort analysis, which defines a study population after starting treatments but includes follow-up data before the treatments, would make effect estimates vulnerable to selection bias [3, 23]. The results would not be only biased for some target parameters, but also blur the target parameters in real world settings. For any database, sample selection strategies to define "time-zero" at meeting the hypothetical eligibility criteria for a "target trial" [13, 24] will be required to clearly define and estimate the target parameters.

2.4 Confounding

As introduced earlier, causal inference methodology defines treatment effects through counterfactual comparison on the same individual or group. As an example, suppose that we are interested in the preventive effect of postoperative antimicrobial treatment A compared to no treatment on the rate of surgical site infections for patients who underwent orthopedic surgery. Let R_G denote the conditional risk of infection in subgroup G and R_G^T denote the counterfactual risk if subgroup G had received treatment T. Then, the effect of treatment A compared with no-treatment may be defined as $R_A - R_A^0$, $R_0^A - R_0$, or $R_{A+0}^A - R_{A+0}^0$ ("0" means no-treatment and subscript "A + 0" means a union of treatment groups A and 0), depending on the population where the effect is defined. The study population is described as being *comparable* or *free from confounding* with respect to effect of treatment A (relative to no-treatment) if the following conditions are met:

(1)   Had patients who have received treatment A been instead operated with no treatment, their infection risk (R_A^0) would be the same as the observed infection risk among the patients with no treatment (R_0), *and*

(2)   Had patients who have received no treatment been instead operated with treatment A, their infection risk (R_0^A) would be the same as the observed infection rate among the patients with treatment A (R_A).

If the population is comparable, the treatment effects defined above coincide with each other and are estimated by the sample risk difference $R_A - R_0$.

In an RCT, randomization splits patients into two groups irrespective of their prognostic factors, assuring comparability. Unfortunately, comparability is not generally achieved in an RWD, as patients in clinical practice receive particular treatments because of their clinical indications. The situation represents a *confounding-by-indication*, widely known in pharmacoepidemiology [25].

There is a statistical solution to confounding in a non-randomized setting: stratification. Stratification creates strata in which patients' prognostic factors are identical. If we can stratify data finely enough, patients with treatment A and no-treatment may be comparable within each stratum. For example, age, sex, and American Society of Anesthesiologists (ASA) physical status are known risk factors for surgical site infection [26]. After stratification, there is little or no variation of these factors within strata (e.g., 40–45-year old men with an ASA physical status of 3). By stratifying on more and more risk factors, we can expect the following comparability is likely approximated: (1) $R_{s,A}^0 = R_{s,0}$ and (2) $R_{s,0}^A = R_{s,A}$ for every stratum $s = 1,..., S$ with stratum-size of $N_{s,A+0} = N_{s,A} + N_{s,0}$ (subscript "s,G" means an intersection of stratum s and treatment G). The set of variables that are sufficient to make treatment groups conditionally comparable within strata are *confounders*. If we believe that data can be stratified based on confounders from available information, we make a *no-unmeasured confounders* assumption in the analysis. Under this assumption, counterfactual risks are $R_A^0 = \sum_{s=1}^{S} N_{s,A} R_{s,0}/N_A$, $R_0^A = \sum_{s=1}^{S} N_{s,0} R_{s,A}/N_0$, $R_{A+0}^A = \sum_{s=1}^{S} N_{s,A+0} R_{s,A}/N_{A+0}$, and $R_{A+0}^0 = \sum_{s=1}^{S} N_{s,A+0} R_{s,0}/N_{A+0}$, all of which are estimated by the observed risks within strata. These effects are different in the presence of confounding, depending on the choice of the set of "standardizing weights" for stratified risks: the weight $N_{s,A}/N_A$ targets population actually operated with treatment A (i.e., $R_A - R_A^0$), $N_{s,0}/N_0$ targets the population actually operated without treatment (i.e., $R_0^A - R_0$), and $N_{s,A+0}/N_{A+0}$ targets the combined population (i.e., $R_{A+0}^A - R_{A+0}^0$). The *target population* of causal effects on which a counterfactual comparison is defined may be internal, as above, or external by selecting the weights. If the external population is targeted, generalizability and transportability are considered as the counterfactual prediction issues [27, 28].

In general, confounders are not a unique set of variables, as they are sufficient sets that bring approximate comparability between treatment groups. A causal directed acyclic graph would help analysts to select appropriate confounders to be stratified [10, 23], but the no-unmeasured confounders assumption is never assured from the data [1, 3, 9]. Intermediate (mediating) variables between treatment and outcome should not generally be included in the confounders unless the analytical objective is to separate treatment effect into direct and indirect effects; however, that analysis requires specialized statistical techniques [29].

Formal causal inference theory implicitly requires additional assumptions: no-interference (outcome is not affected by other patients' treatment values), no-multiple versions of treatments (a counterfactual risk under one treatment is not dependent on how to administer the treatment, or there is no choice of administration methods), and positivity (all strata include both patients with treatment A and patients with no-treatment). In our example, no-interference may be likely but no-multiple version would be violated by the different durations of postoperative treatment or surgical skills/circumstances [30]. Only positivity can be empirically checked from data at hand.

3 Statistical Methods to Handle the Confounding Bias

3.1 Regression and Propensity Score Modeling to Approximate Stratified Analysis

We usually have to stratify data on the moderate to large number of confounders. Even for only 10 binary confounders, there should be $2^{10} = 1024$ strata and the stratum-number will increase exponentially with increasing number of confounders and their categories. Thus, necessary strata will be more than sample sizes of *any* datasets in reality, even with sample size of millions or billions. RWD also suffer from this *curse of dimensionality*, because confounding-by-indication in real-world settings requires too many variables to stratify.

We use statistical models to approximately estimate stratum-specific risks $R_{s,0}$ and $R_{s,A}$ from a limited amount of data. If outcome is binary, logistic models are conventionally fitted, such as $R_{s,A} = \{1 + \exp(-\beta_0 - \beta_1 - \boldsymbol{\beta_2}\boldsymbol{C_s})\}^{-1}$ and $R_{s,0} = \{1 + \exp(-\beta_0 - \boldsymbol{\beta_2}\boldsymbol{C_s})\}^{-1}$, where $\boldsymbol{C_s}$ is a set of confounders' values that correspond to stratum s. After fitting the models, the estimate of $(\beta_0, \beta_1, \boldsymbol{\beta_2})$ can be used to predict $R_{s,0}$ and $R_{s,A}$ for each combination of $\boldsymbol{C_s}$. Equivalently, if each patient $i = 1,\ldots,$ N_{A+0} has distinct value $\boldsymbol{C_i}$, the logistic models $R_{i,A} = \{1 + \exp(-\beta_0 - \beta_1 - \boldsymbol{\beta_2}\,\boldsymbol{C_i})\}^{-1}$ and $R_{i,0} = \{1 + \exp(-\beta_0 - \boldsymbol{\beta_2}\,\boldsymbol{C_i})\}^{-1}$ can predict $R_{i,0}$ and $R_{i,A}$ irrespective of actual treatment status. Averaging and comparing the predicted risks $R_{i,0}$ and $R_{i,A}$ among a target population provides the model-based *regression-standardized* estimates of the treatment effect. For example, the estimator targeting the combined population is $\sum_{i=1}^{N_{A+0}} R_{i,A}/N_{A+0} - \sum_{i=1}^{N_{A+0}} R_{i,0}/N_{A+0}$.

Another approach to approximate stratified analysis is to model the *propensity score*, which is the conditional probability of treatment A within stratum s, denoted by p_s. Within strata $s = 1,\ldots, S$, the confounders are identical between treatment groups; within strata of p_s, while the confounders are not necessarily identical between the groups, but the "balancing property" of propensity score assures the same confounder distribution between the groups [31]. As a result, standardized p_s-stratified risks following stratification on p_s is identical to confounder-stratum s-standardized estimates from $R_{s,0}$ and $R_{s,A}$. In practice, we need to rely on statistical models, such as $p_i = \{1 + \exp(-\alpha_0 - \alpha_1 C_i)\}^{-1}$ for patient i, to approximate p_s; stratification/regression adjustment for p_i, matching on p_i, and inverse probability weighting can be used to estimate treatment effects corresponding to target populations of interest, but these methods generally provide different estimates.

Both approaches assume the correct model specification. To enhance the approximation, higher-order terms, such as product-terms between treatment and confounders or between confounders and polynomial terms for continuous variables may have to be included. Also, we are not interested in the regression parameters themselves (i.e., β's and α's). Rather, the models are imposed to estimate $(R_{s,0}, R_{s,A})$ or p_s to restore the stratified analysis from sparse data with innumerable confounders [9]. Although regression parameters are interpretable as individual- or stratum-specific effects, they may be quite different from marginalized effects in target populations, especially for logistic models [32]. Finally, we do not need to select either a regression model or a propensity-score model. If we correctly combine them into a single estimator, we can protect estimates against model misspecification through *doubly robust estimators*, which provide asymptotically unbiased estimates of exposure effects in target populations if at least one model of regression or propensity score is correctly specified [33–38].

3.2 Methods for Time-Dependent Confounding in Sustained/Repeated Treatments

Unlike our previous example of postoperative antimicrobial treatments, most treatments/interventions in RWD are sustained or repeated throughout the follow-up period. The effects of such treatments should be defined by the "sequence" of treatments [12, 13, 16, 24, 39].

However, an analytical difficulty arises for confounding with sustained treatments sequences. In clinical practice, treatments are usually administered according to a patient's clinical conditions, which introduce confounding-by-indication; the treatments affect the patient's condition; and again the treatment decision is made based on the changed patient's condition. The cycle will continue. The patient's changing condition is an example of *time-dependent confounders*, which are measured after the start of treatment sequences and need to be stratified to make subsequent treatment subgroups comparable [39]. Even if we measure all possible time-dependent confounders, however, stratification cannot purge bias when the time-dependent confounders are affected by previous treatments [9, 12, 39].

Modern causal inference methodology has developed methods to estimate the effects of sustained treatments under the assumption of the sequential conditional comparability: within subgroups stratified on the history of past treatments and time-dependent and -independent confounders, treatment groups at every time-point are comparable, or the no-unmeasured confounders for sequential treatments during the follow-up [9, 12, 39]. Standardization via stratified analysis has been extended to the *g-formula* [10, 12] but fine stratification for measured confounders at every time-point almost always requires parametric modeling. Parametric g-formula approximates the g-formula through regression modeling of stratum-specific risks and confounder distributions at each time-point given past treatment and confounder histories, and through Monte-Carlo evaluation of fitted models. Alternatively, *structural nested models* impose functional restrictions on the direct-effect of treatment at each time-point by subtracting subsequent treatment effects. The effects are conditional on past treatment and confounder histories, and are sequentially estimated by the g-estimation procedure via propensity score at each time-point. *Marginal structural models* are another class of models for causal contrast, which model counterfactual risks under different sequence of treatments; inverse probability weighting via the propensity scores for sequential treatments can be directly applicable for estimation. These methods are applicable to not only the effects of predetermined treatment sequences, but also the effects of dynamic treatment regimes, which are determined following changes in patient's clinical conditions [13, 39–42].

3.3 Applicability of Machine-Learning and Quasi-Experimental Techniques

While it is accepted that machine learning techniques, such as random forest or neural network, are powerful means of prediction with modern computing circumstances [6], application of these methods to causal inference is still in progress in the literature. Since the objective of modeling in causal inference is to approximate stratified analysis (for single time-point treatments) or g-formula/structural models (for sustained treatments), machine learning methods would greatly improve our toolkits for causal inference as well [43]. Even for doubly-robust estimators, at least one model must be specified correctly and sparse-data bias arises if extreme risks or propensity scores exist in some strata. However, it must be remembered that the machine learning approach itself cannot serve as a framework for causal formulation of treatment effects; a counterfactual formulation and machine learning will play complementary roles in causal analysis with RWD.

If there are unmeasured confounders, all the aforementioned methods will be biased. In some situations, quasi-experimental techniques like instrumental variable estimation [11, 44], regression discontinuity [44, 45], and interrupted time series [46] are available to de-confound the treatment effects without confounder information. However, this comes with a price; they must rely on alternate unverifiable assumptions instead of no-unmeasured confounders, including unrealistically simplified causal models. Although they serve as sensitivity analyses under different assumptions for different aspects of datasets, there is no savior in RWD analysis.

4　Summary

We are living in the era of RWD, which provides a great opportunity to establish effective treatment strategies for future patients. However, RWD databases have their own pitfalls compared to traditional study designs like RCTs or epidemiological designs. Moreover, the massive amount of data would produce p-values that are too small, especially when using simple statistical models [47]. Thus, we have to be responsible to appraise such illusionary conclusions from RWD databases, keeping our eyes open to the bias sources introduced in this chapter. While sophisticated analytical techniques would improve the quality of inference, they cannot overcome the quality of data. Setting strict eligibility criteria depending on the research questions may be required in RWD analysis to avoid bias from incomplete data [13, 24].

It may be helpful to follow guidelines, such as RECORD for observational routinely collected health data [48]. Although designed for planned RCTs, CONSORT would apply RWD studies that attempt to emulate hypothetical RCTs in RWD databases for causal analysis [47, 49].

References

1. Rothman KJ, Greenland S, Lash TL (eds) (2008) Modern epidemiology. 3rd edn. Lippincott Williams and Wilkins
2. Gatsonis C, Morton SC (eds) (2017) Methods in comparative effectiveness research. Chapman and Hall/CRC
3. Greenland S, VanderWeele TJ (2015) Validity and bias in epidemiological research. In: Detels R, Gulliford M, Karim QA, Tan CC (eds) Oxford textbook of global public health, 6th edn. Oxford University Press
4. Harrell F (2001) Regression modeling strategies. Springer
5. Steyerberg EW (2008) Statistical models for prediction. Springer
6. Hastie T, Tibshirani R, Friedman J (2008) The elements of statistical learning: data mining, inference, and prediction, 2nd edn. Springer
7. Greenland S, Mansournia MA, Altman DG (2016) Sparse data bias: a problem hiding in plain sight. Br Med J 353:i1981
8. Vickers AJ, Elkin EB (2006) Decision curve analysis: a novel method for evaluating prediction models. Med Decis Mak 26:565–574
9. Hernán MA, Robins JM (2020) Causal inference: what if. Chapman and Hall/CRC

10. Pearl J (2009) Causality: models reasoning, and inference, 2nd edn. Cambridge University Press
11. Imbens GW, Rubin DB (2015) Causal inference in statistics, social, and biomedical sciences: an introduction. Cambridge University Press
12. Robins JM (1986) A new approach to causal inference in mortality studies with sustained exposure periods-application to control of the healthy worker survivor effect. Comput Math Appl 14:1393–1512
13. Hernán MA, Robins JM (2016) Using big data to emulate a target trial when a randomized trial is not available. Am J Epidemiol 183:758–764
14. Molenberghs G, Fitzmaurice G, Kenward MG, Tsiatis A, Verbeke G (eds) (2014) Handbook of missing data methodology. Chapman and Hall/CRC
15. Klein JP, Moeschberger ML (2005) Survival analysis: techniques for censored and truncated data, 2nd edn. Springer
16. Robins JM, Finkelstein DM (2000) Correcting for noncompliance and dependent censoring in an AIDS clinical trial with inverse probability of censoring weighted (IPCW) log-rank tests. Biometrics 56:779–788
17. Uno H, Cai T, Tian L, Wei LJ (2007) Evaluating prediction rules for t-year survivors with censored regression models. J Am Stat Assoc 102:527–537
18. Allison DA (2010) Survival analysis using SAS: a practical guide. SAS Institute
19. Yasunaga H, Matsui H, Horiguchi H, Fushimi K, Matsuda S (2014) Application of the diagnosis procedure combination (DPC) data to clinical studies.(in Japanese). J UOEH 36:191–197
20. Matsuda S, Fujimori K, Kuwabara K, Ishikawa KB, Fushimi K (2011) Diagnosis procedure combination as an infrastructure for the clinical study. Asian Pac J Dis Manag 5:81–87
21. Greenland S (1980) The effect of misclassification in the presence of covariates. Am J Epidemiol 112:564–569
22. Rothman KJ, Gallacher JE, Hatch EE (2013) Why representativeness should be avoided. Int J Epidemiol 42:1012–1014
23. Hernán MA, Hernández-Díaz S, Robins JM (2004) A structural approach to selection bias. Epidemiology 15:615–625
24. Danaei G, García Rodríguez LA, Cantero OF, Logan RW, Hernán MA (2018) Electronic medical records can be used to emulate target trials of sustained treatment strategies. J Clin Epidemiol 96:12–22
25. Joffe MM (2000) Confounding by indication: the case of calcium channel blockers. Pharmacoepidemioly Drug Saf 9:37–41
26. Ban KA, Minei JP, Laronga C et al (2017) American college of surgeons and surgical infection society: surgical site infection guidelines, 2016 UPDATE. J Am Coll Surg 224:59–74
27. Cole SR, Stuart EA (2010) Generalizing evidence from randomized clinical trials to target populations: the ACTG 320 trial. Am J Epidemiol 172:107–115
28. Bareinboim E, Pearl J (2013) A general algorithm for deciding transportability of experimental results. J Causal Inference 1:107–134
29. Vanderweele TJ (2015) Explanation in causal inference: methods for mediation and interaction. Oxford University Press
30. Berríos-Torres SI, Umscheid CA, Bratzler DW, Leas B, Stone EC, Kelz RR, Reinke CE, Morgan S, Solomkin JS, Mazuski JE, Dellinger EP, Itani KMF, Berbari EF, Segreti J, Parvizi J, Blanchard J, Allen G, Kluytmans JAJW, Donlan R, Schecter WP (2017) Healthcare infection control practices advisory committee. Centers for disease control and prevention guideline for the prevention of surgical site infection. JAMA Surg 152:784–91
31. Rosembaum PR, Rubin DB (1983) The central role of the propensity score in observational studies for causal effects. Biometrika 70:41–55
32. Greenland S, Robins JM, Pearl J (1999) Confounding and collapsibility in causal inference. Stat Sci 14:29–46
33. Bang H, Robins JM (2005) Doubly robust estimation in missing data and causal inference models. Biometrics 61:962–973

34. Li L, Greene T (2013) A weighting analogue to pair matching in propensity score analysis. Int J Biostat 9:215–234
35. Shinozaki T, Matsuyama Y (2015) Doubly robust estimation of standardized risk difference and ratio in the exposed population. Epidemiology 26:873–877
36. Hattori S, Henmi M (2014) Stratified doubly robust estimators for the average causal effect. Biometrics 70:270–277
37. Vansteelandt S, Daniel RM (2014) On regression adjustment for the propensity score. Stat Med 33:4053–4072
38. Shinozaki T, Nojima M (2019) Misuse of regression adjustment for additional confounders following insufficient propensity-score balancing. Epidemiology 30:541–548
39. Robins JM, Hernán MA (2009) Estimation of the causal effects of time-varying exposures. In: Fitzmaurice G, Davidian M, Verbeke G, Molenberghs G (eds) Longitudinal data analysis. Chapman and Hall/CRC Press
40. Young JG, Cain LE, Robins JM, O'Reilly EJ, Hernán MA (2011) Comparative effectiveness of dynamic treatment regimes: an application of the parametric g-formula. Stat Biosc 3:119–143
41. Schulte PJ, Tsiatis AA, Laber EB, Davidian M (2014) Q- and A-learning methods for estimating optimal dynamic treatment regimes. Stat Sci 29:640–661
42. Hernán MA, Lanoy E, Costagliola D, Robins JM (2006) Comparison of dynamic treatment regimes via inverse probability weighting. Basic Clin Pharmacol Toxicol 98:237–242
43. van der Laan MJ, Rose S (2018) Targeted learning in data science. Springer
44. Angrist JD, Pischke JS (2008) Mostly harmless econometrics: an empiricist's companion. Princeton University Press
45. Oldenburg CE, Moscoe E, Bärnighausen T (2016) Regression discontinuity for causal effect estimation in epidemiology. Curr Epidemiol Rep 3:233–241
46. Bernal JL, Cummins S, Gasparrini A (2017) Interrupted time series regression for the evaluation of public health interventions: a tutorial. Int J Epidemiol 46:348–355
47. Hernán MA (2011) With great data comes great responsibility: publishing comparative effectiveness research in epidemiology. Epidemiology 22:290–291
48. Benchimol EI, Smeeth L, Guttmann A, Harron K, Moher D, Petersen I, Sørensen HT, von Elm E, Langan SM (2015) Record working committee. The reporting of studies conducted using observational routinely-collected health data (RECORD) statement. PLoS Med 12:e1001885
49. Schulz KF, Altman DG, Moher D (2010) The CONSORT group. CONSORT 2010 statement: updated guidelines for reporting parallel group randomised trials. Ann Intern Med 152:726–32

Problems in Japanese Real-World Medical Data Analyses

Shoji Tokunaga

1 A Mini Review of Studies Analyzing the Japanese Real-World Medical Data (RWMD)

1.1 The Method of Mini Review

Studies on Japanese RWMD were reviewed to extract problems in the Japanese RWMD analyses. The main reviewed items were the reported study design, database, statistical methods, and limitations. Although the review process was not systematic, I believe that this review could unveil some latent problems in the Japanese RWMD analyses.

The studies published on line in 2018 were searched on PUBMED with the keywords "Japan" and "database." The studies that retrospectively analyzed the Japanese RWMD in the fields of observational epidemiology and pharmacoepidemiology were collected. Prospective cohort studies with predetermined study objectives and a set of variables addressing the objectives were excluded.

As a result, 90 published studies were selected for the analysis (see Appendix for the list of papers). The studies were classified into eight categories according to the study theme. The database, statistical method, statistical software, and limitations listed in the studies were tabulated based on the theme category.

The eight theme categories were (1) "Comparison," comparison of the clinical benefit or risk between medications, patient characteristics, or conditions (this category include prediction study on the prognosis or treatment effect), (2) "Description," the descriptive epidemiology with few or no comparison among groups of subjects, (3) "Adverse events," analyses on the adverse events of medications, (4) "Cost," analyses on the costs of treatment and health care, (5) "Methodology," development

S. Tokunaga (✉)
Medical Information Center, Kyushu University Hospital, Fukuoka, Japan
e-mail: tokunaga.shoji.642@m.kyushu-u.ac.jp

© The Author(s), under exclusive license to Springer Nature Singapore Pte Ltd. 2022   89
N. Nakashima (ed.), *Epidemiologic Research on Real-World Medical Data in Japan*,
SpringerBriefs for Data Scientists and Innovators 2,
https://doi.org/10.1007/978-981-19-1622-9_13

of the methodology of statistical analyses, (6) "Verification of the data," (7) "Drug repositioning" and (8) "Drug–drug interaction."

Table 1 summarizes the databases frequently used in the studies, which were as follows: Diagnosis Procedure Combination (DPC), Japan Medical Data Center (JMDC), Medical Data Vision (MDV), Japanese Adverse Drug Event Report (JADER), National Database of Health Insurance Claims and Specific Health Checkups of Japan (NDB), National Clinical Database (NCD), and nationwide Japan Adult Cardiovascular Surgery Database (JCVSD) which is based on NCD. The names of less-frequently used DBs as well as the set of DBs used as a combination are shown as footnotes of Table 1.

Statistical methods used in the studies were categorized into (1) "Univariate analysis," (2) "Multivariable analysis," (3) "Complex statistical model," (4) "Propensity score analysis," (5) "Instrumental variable analysis," (6) "Adverse events analysis," and (7) "Other methods." The statistical analyses were also performed to determine whether the analysis was conducted with "Consideration on clustering of patients" and whether "Multiple imputation for missing data" was done. Other categories of statistical analyses were "Estimations" and "Development and validation of the model." The footnotes of Table 2 describe the example of each category; Table 3 shows the statistical software used in the studies.

Limitations pointed out by the authors of the studies were classified into (1) Data: quality of data limited further analyses. It was subclassified into "Unavailable variables," "Insufficient accuracy," "Lack of details," "Poor or no validation," "Misclassification," and "Miscellaneous;" (2) "Selection bias," the bias caused by patient's age, employment status and health condition, hospital characteristics etc.; (3) "Follow-up," unable to get long-term outcomes owing to incomplete follow-up caused by emigration or hospital discharge; (4) "Information bias," the bias inherent to spontaneous reporting system or the lack of a part of data such as uninsured treatment claims;" (5) "Subjects," subjects consisted of heterogeneous populations, the total number of unavailable patients owing to spontaneous reporting system etc.; (6) "Statistical analysis," a large number of unmatched patients at propensity score matching, very small number of patients for subgroup analysis, the confounding factors could not be fully adjusted at the analysis, the unit of analysis was a admission rather than a patient, etc.; (7) "Database" impossible to link DBs; and (8) "Representativeness," too few patients.

1.2 The Results of the Review

The number of studies tabulated by the theme and database were shown in Table 1. The papers classified as "Comparison" were most abundant (55.6%), followed by "Description" (22.2%), "Adverse events" (8.9%), "Cost" (6.7%), and "Methodology" (3.3%). The number of studies belonged to "Verification of the data," "Drug repositioning," and "Drug–drug interaction" were one (1.1%) for each.

Table 1 Number of studies categorized by the theme and database used

Category of the theme[b]	Number of papers	Database[a] DPC	JMDC	MDV	JADER	NDB	NCD	JCVSD	Linked DBs[c]	Other databases[d]
Comparison	50 (55.6)[e]	18 (36.0)[f]	9 (18.0)	4 (8.0)	1 (2.0)	5 (10.0)	5 (10.0)	1 (2.0)	4 (8.0)	3 (6.0)
Description	20 (22.2)	4 (20.0)	5 (25.0)	5 (25.0)	0	2 (10.0)	1 (5.0)	0	1 (5.0)	2 (10.0)
Adverse events	8 (8.9)	0	0	0	8 (100.0)	0	0	0	0	0
Cost	6 (6.7)	0	3 (50.0)	2 (33.3)	0	0	0	0	1 (16.7)	0
Methodology	3 (3.3)	0	0	0	2 (66.7)	1 (33.3)	0	0	0	0
Verification of the data	1 (1.1)	0	0	0	0	0	1 (100.0)	0	0	0
Drug repositioning	1 (1.1)	0	1 (100.0)	0	0	0	0	0	0	0
Drug–drug interaction	1 (1.1)	0	0	0	1 (100.0)	0	0	0	0	0
Total	90 (100.0)	22 (24.4)	18 (20.0)	11 (12.2)	12 (13.3)	8 (8.9)	7 (7.8)	1 (1.1)	6 (6.7)	5 (5.6)

[a] DPC, Diagnosis Procedure Combination; JMDC, the Japan Medical Data Center database; MDV, the Medical Data Vision database; JADER, the Japanese Adverse Drug Event Report; NDB, the National Database of Health Insurance Claims and Specific Health Checkups of Japan; NCD, the National Clinical Database; JCVSD, the nationwide Japan Adult Cardiovascular Surgery Database (based on NCD)

[b] See text for the definition

[c] JADER + medical chart review, NDB + "630 survey" (the nationwide study collecting information on new psychiatric admissions), the Specific Health Check-Up for all inhabitants of Japan + the database containing death certificates for all deaths, DPC + Hospital-based cancer registries, JMDC + MDV, NDB + the Survey of Long-Term Care Benefit Expenditures + an informal care time survey for informal care cost

[d] The Fukuoka Prefecture Wide-Area Association of Latter-Stage Elderly Healthcare, a Japanese electronic medical records database (three million patients from more than 60 hospitals), the Japan Society of Gastroenterological Surgery registry on the NCD, four pharmacy claims datasets provided by the nationwide pharmacy chains, and the Annual Report on Health and Welfare in Nagoya City

[e] Number of studies (column%)

[f] Number of studies (% of the papers belong to the theme category)

Table 2 Statistical methods used in the studies

Theme (number of papers)	Statistical methods[a]										
	Univariate analysis	Multivariable analysis	Complex statistical models	Propensity score analysis	Instrumental variable analysis	Adverse events analysis	Other statistical methods	Consideration on clustering of patients	Multiple imputation for missing data	Estimations	Development and validation of database
Comparison (50)	20 (40.0)[b]	31 (62.0)	4 (8.0)	17 (34.0)	4 (8.0)	2 (4.0)	1 (2.0)	7 (14.0)	4 (8.0)	5 (10.0)	3 (6.0)
Description (20)	14 (70.0)	2 (10.0)	0	0	0	0	0	0	0	1 (5.0)	0
Adverse events (8)	2 (25.0)	1 (12.5)	0	0	0	7 (87.5)	0	0	0	0	0
Cost (6)	3 (50.0)	4 (66.7)	0	0	0	0	1 (16.7)	0	0	0	0
Methodology (3)	0	0	0	0	0	0	1 (33.3)	0	0	2 (66.7)	0
Verification of the data (1)	0	0	0	0	0	0	0	0	0	1 (100.0)	0
Dug repositioning (1)	0	0	0	0	0	1 (100.0)	0	0	0	0	0
Drug-drug interaction (1)	0	0	0	0	0	1 (100.0)	0	0	0	0	0

(continued)

Table 2 (continued)

| Theme (number of papers) | Statistical methods[a] | | | | | | | | | | | |
	Univariate analysis	Multivariable analysis	Complex statistical models	Propensity score analysis	Instrumental variable analysis	Adverse events analysis	Other statistical methods	Consideration on clustering of patients	Multiple imputation for missing data	Estimations	Development and validation of database
Total (90)	39 (43.3)	38 (42.2)	4 (4.4)	17 (18.9)	4 (4.4)	11 (12.2)	3 (3.3)	7 (7.8)	4 (4.4)	9 (10.0)	3 (3.3)

[a] Classification (examples) of the statistical methods; "Univariate analysis" (Chi-square test, t-test, log-rank test, simple regression analysis including linear, logistic and Cox models), "Multivariable analysis" (mainly of logistic and Cox models), "Complex statistical model" (marginal structural model, mixed effects model, adjustment for time-dependent confounders by inverse probability treatment weights), "Propensity score analysis", "Instrumental variable analysis", "Adverse events analysis" (the reporting odds ratio, the proportional reporting ratio, the association rule mining, the modified existing association rule mining, the shape parameter beta of the Weibull distribution), and "Other methods" ('probabilistic modeling', the association rule mining, the cost minimization analysis), "Consideration on clustering of patients" (clustering within hospitals and/or doctors), "Multiple imputation for missing data" (the pooled estimates from the multiply imputed datasets according to the statistical model), "Estimations" (the area under the ROC curve, E values which show what size an unmeasured confounder would have to be to nullify its effect, the net reclassification index, the integrated discrimination index, violin plots, box plots, spline curves, sensitivity, specificity) and the "Development and validation of the model" (the cohort was randomly divided into an 80% model development cohort and a 20% testing cohort)

[b] Number of papers (% in the total number of papers belonging to the theme category)

Table 3 Statistical software used in the studies

Theme (number of papers)	SAS	Stata	R	SPSS	JMP	Visual Mining Studio	Combinations[a]	Not shown
Comparison (50)	9 (18.0)[b]	15 (30.0)	7 (14.0)	8 (16.0)	2 (4.0)	0	6 (12.0)	3 (6.0)
Description (20)	10 (50.0)	1 (5.0)	1 (5.0)	1 (5.0)	0	0	2 (10.0)	5 (25.0)
Adverse events (8)	0	0	1 (12.5)	0	3 (37.5)	0	2 (25.0)	2 (25.0)
Cost (6)	4 (66.7)	1 (16.7)	1 (16.7)	0	0	0	0	0
Methodology (3)	0	0	0	0	0	0	2 (66.7)	1 (33.3)
Verification of the data (1)	0	0	0	0	1 (100.0)	0	0	0
Dug repositioning (1)	0	0	0	0	0	1 (100.0)	0	0
Drug–drug interaction (1)	1 (100.0)	0	0	0	0	0	0	0
Total (90)	24 (26.7)	17 (18.9)	10 (11.1)	9 (10.0)	6 (6.7)	1 (1.1)	12 (13.3)	11 (12.2)

[a] Combinations of software (number of papers); SAS + SPSS (2), Stata + SPSS (3), Stata + JMP (1), R + SPSS (1), R + JMP (2), Visual Mining Studio + JMP (3)
[b] Number of studies (row%)

The most frequently used databases were DPC (24.4%), JMDC (20.0%), MDV (12.2%), and JADER (13.3%). The commercially available JMDC and MDV were totaled to nearly one third (32.2%) of the selected studies. Other databases are JADER, NDB, NCD, and JCVSD which may be included in NCD. Only six studies (6.7%) used the linked databases.

The distribution of the analyzed databases was different according to the study theme. Among the studies of "Comparison," DPC was most frequently used (36.0%). Studies in the "Description" and "Cost" tended to use commercially available databases (50.0% and 83.3%, respectively). "Adverse events" analyses were conducted solely on JADER.

Table 2 shows the statistical methods used by the studies. Over 40% of studies used "Univariate and/or Multivariable analyses," followed by "Propensity score analysis" (18.9%). Studies in different theme categories tend to use different statistical methods. In the category "Comparison," 62.0% of studies used "Multivariable analysis," and 34% used "Propensity score analysis." Use of "Complex statistical models," "Propensity score analysis," "Instrumental variable analysis," "Consideration of clustering," "Multiple imputation for missing data," and "Development and validation of the database" were used only in the studies in "Comparison." The studies classified to "Description" heavily relied on "Univariate analysis." "Adverse events (AEs) analysis" frequently used reporting odds ratio and other statistical methods specific to AE analysis such as the proportional reporting ratio (PRR), signal value (=ln(PRR) + ln(χ^2)), shape parameter beta of the Weibull distribution and association rule mining. It is suggested that the use of these statistical methods depends on the spontaneous reporting system of JADER.

Table 3 shows the statistical software used in the studies. The most frequently used software was SAS (26.7%), followed by Stata (18.9%), R (11.1%), SPSS (10.0%), JMP (6.7%), and Visual Mining Studio (1.1%). Among the studies on "Comparison," Stata (30.0%) was used the most, followed by SAS (18.0%), SPSS (16.0%), and R (14.0%). Half of the studies in "Description" used SAS. Considering the 70% of studies in "Description" conducted univariate analysis, SAS might be used mainly for the data management wasting its high functionality of statistical analysis.

Limitations pointed out in the studies were categorized and shown in Table 4. Limitations in the data quality were most frequently appeared (77.8%), followed by "Selection bias" (26.7%) and "Follow-up" (18.9%). Examples of the latter two categories of limitations were: (1) the population of the commercial database is limited to the insures of the contracted healthcare societies (JMDC), (2) NDB does not include unemployed individuals and those over the age of 65 years are scarce, (3) Some hospitals, especially small scale hospitals, do not introduce DPC, and (4) DPC cannot follow the patients moving to other centers or after hospital discharge and its follow-up duration was insufficient to access the long-term outcomes.

Other categories listed were "Information bias" (15.6%), "Subjects" (11.1%), "Statistical analysis" (10.0%), and the rest (2.2%). Among on the studies in "Comparison," "Description," "Adverse events," and "Cost," the limitations due to "Data" especially the subclassified "Unavailable variables" were frequently appeared. The limitation might be partly due to the retrospective nature of Japanese RWMD analyses, of which the set of variables was not pre-planned for particular studies.

In the studies in "Adverse events," limitations due to "Information bias" and "Subjects" were frequently stated, possibly by due to the data collected from spontaneous reporting system, hence, the subjects did not include the healthy individuals.

2 The Problems in Japanese RWMD Analyses

The review showed several existing problems in the Japanese RWMD analyses. First, the lack of the linkage among the databases could possibly result in a part of limitations in analyses. Authors of the review papers often reported a lack of variables and insufficient follow-up as the limitations in analyses. Some risk factors or variables needed to adjust confounding factors might not be included in the dataset. Long-term outcomes could not be confirmed as it was impossible to identify whether the patient had recovered, transferred to another care institution or died in most databases, including DPC, NCD, and NBD. If the databases could be linked together, the insufficient set of variables may be fill up. The exact linkage among the databases,

Table 4 Limitations pointed out by the authors of the papers

Theme (number of papers)	Data	Subclassification of "Data"					Misc.	Selection bias	Followup	Information bias	Subjects	Statistical analysis	Database	Representativeness
		Unavailable variables	Insufficient accuracy	Lack of details	Poor or no validation	Misclassification								
Comparison (50)	39 (78.0)[a]	28 (56.0)	9 (18.0)	11 (22.0)	4 (8.0)	3 (6.0)	4 (8.0)	10 (20.0)	12 (24.0)	3 (6.0)	2 (4.0)	8 (16.0)	1 (2.0)	0
Description (20)	15 (75.0)	11 (55.0)	6 (30.0)	3 (15.0)	1 (5.0)	1 (5.0)	2 (10.0)	11 (55.0)	4 (20.0)	1 (5.0)	2 (10.0)	0	0	0
Adverse events (8)	8 (100)	7 (87.5)	3 (37.5)	1 (12.5)	0	0	2 (25.0)	0	0	5 (62.5)	5 (62.5)	0	0	0
Cost (6)	5 (83.3)	4 (66.7)	1 (16.7)	0	1 (16.7)	0	0	3 (50.0)	1 (16.7)	1 (16.7)	0	0	0	0
Methodology (3)	1 (33.3)	0	1 (33.3)	0	0	0	0	0	0	3 (100)	0	0	0	0
Verification of the data (1)	0	0	0	0	0	0	0	0	0	0	1 (100)	0	0	1 (100)
Drug repositioning (1)	1 (100)	0	1 (100)	0	0	0	0	0	0	1 (100)	0	0	0	0
Drug-drug interaction (1)	1 (100)	1 (100)	0	0	0	0	0	0	0	0	0	1 (100)	0	0
Total (90)	70 (77.8)	51 (56.7)	21 (23.3)	15 (16.7)	6 (6.7)	4 (4.4)	8 (8.9)	24 (26.7)	17 (18.9)	14 (15.6)	10 (11.1)	9 (10.0)	1 (1.1)	1 (1.1)

[a] Number of studies (row%)

however, may not be possible in Japan due to the anonymous databases and a lack of common identification code among databases. If these databases had been registered by a common code, the databases could be linked together, and the variables could complement each other or the patients may be followed up beyond the period that a single database could follow.

Second, the authors of review papers have pointed out insufficient accuracy and misclassification as the possible problems in the data. There were several studies dealing with the data validity. In the review papers, a paper examined the data validity in NCD and showed a high accuracy of data input [1]. In 2017, the validity of diagnoses and procedure records in the DPC data and laboratory results in the SS-MIX data were reported to be generally high [2]. Although more validity studies are needed, the data validity might be sufficient for the objectives of the analyses.

Third, the quality of statistical analyses was low in several studies. Although a number of studies have adjusted confounding factors by the multivariable models and sometimes more complex models, some authors used only univariate analysis and neglected the confounding factors. Also, the "Considering the clustering of patients" and "multiple imputation for missing data" were applied in limited studies. Only one study conducted analysis with competing risk model based on the Fine and Gray's model, and the models that included time-dependent variable were used only in two studies. Thus a small proportion of Japanese studies in RWMD analysis adopted a high level of statistical methodology. However, quite a few studies seem to be left behind the recent advances in statistical methodology.

Fourth, the SPSS and JMP which are often operated with "point-and-click" interface were used as a statistical software in some studies. If the analyses were conducted manually without saving the scripts of the data management and final analysis, the results of the analyses will be difficult to be reproduced and examined its validity by other researchers.

Fifth, all AE analyses reviewed in this review paper were relied on JADER, a spontaneous reporting database. The data from spontaneous reporting system is known to be susceptible to reporting bias, including underreporting, notoriety effect (the overall rise in the reported number of topical adverse events) [3], Weber effect (the reported number of AEs tends to decrease with the passage of time after a transient rise immediately after marketing), and masking or cloaking effect (signal scores can be suppressed by a large number of reports, in which the same adverse event is connected with other drugs) [4]. These potential biases in the dataset limit the reliability of the analyses of AEs. Future analyses on the AEs should be conducted with the dataset collected form all subjects with and without AEs, such as MID-NET [5].

Finally, it should be noted that limited studies have analyzed the Japanese RWMD according to the international guidelines. There were only three and two studies analyzed in this review explicitly stated that they reported the results in accordance to the Strengthening the Reporting of Observational Studies (STROBE) statement [6, 7] and the REporting of studies Conducted using Observational Routinely-collected health Data (RECORD) Statement [8, 9], respectively. Although the STROBE statement stated that the reports should explain how the study size was arrived at, only one study referred the sample size justification by statistical calculation.

3 Summary

The findings from the review revealed that the most frequently used databases were DPC (24.4%), JMDC (20.0%), MDV (12.2%), and JADER (13.3%). Over 40% of studies used "Univariate and/or Multivariable analyses," followed by "Propensity score analysis" (18.9%). Most of the studies in "Description" relied on "Univariate analysis." "Adverse events analysis" frequently used methods specific to the analyses of AEs. Among the studies in "Comparison," Stata was used the most (30.0%), followed by and SAS (18.0%), SPSS (16.0%), and R (14.0%) were frequently used. Limitations in the data quality were most frequently pointed out by the authors (77.8%), followed by "Selection bias" (26.7%) and "Follow-up" (18.9%). Limitations and problems in the Japanese RWMD analyses were summarized as follows: (1) a lack of linkage among databases partly due to an absence of the common identification code, (2) low quality statistical analyses in some studies, (3) potential reporting bias of JADER database on which Japanese AE analyses relied on and (4) only a few studies of Japanese RWMD were reported according to the international guidelines, such as STROBE and RECORD statements.

COI There is no COI to declare. There was no financial support from any external organization. This article was written by the author alone.

Appendix

[1] Abe H, Sumitani M, Uchida K, et al (2018) Association between mode of anaesthesia and severe maternal morbidity during admission for scheduled Caesarean delivery: a nationwide population-based study in Japan 2010–2013. Br J Anaesth 120(4):779–789. https://doi.org/10.1016/j.bja.2017.11.101

[2] Akazawa M, Konomura K, Shiroiwa T (2018) Cost-minimization analysis of deep-brain stimulation using national database of Japanese health insurance claims. Neuromodulation 21(6):548–552. https://doi.org/10.1111/ner.12782

[3] Akiyama S, Tanaka E, Cristeau O, Onishi Y, Osuga Y (2018) Treatment patterns and healthcare resource utilization and costs in heavy menstrual bleeding: a Japanese claims database analysis. J Med Econ 21(9):853–860. https://doi.org/10.1080/136 96998.2018.1478300

[4] Anzai T, Takahashi K, Watanabe M (2019) Adverse reaction reports of neuroleptic malignant syndrome induced by atypical antipsychotic agents in the Japanese Adverse Drug Event Report (JADER) database. Psychiatry Clin Neurosci 73(1):27–33. https://doi.org/10.1111/pcn.12793

[5] Arai M, Shirakawa J, Konishi H, Sagawa N, Terauchi Y (2018) Bullous pemphigoid and dipeptidyl peptidase 4 inhibitors: a disproportionality analysis based on the Japanese adverse drug event report database. Diabetes Care 41:e130–e132. https://doi.org/10.2337/dc18-0210

[6] Aso S, Matsui H, Fushimi K, Yasunaga H (2018) Effect of cyclosporine A on mortality after acute exacerbation of idiopathic pulmonary fibrosis. J Thorac Dis 10(9):5275–582. https://doi.org/10.21037/jtd.2018.08.08

[7] Chang CH, Sakaguchi M, Weil J, Verstraeten T (2018) The incidence of medically-attended norovirus gastro-enteritis in Japan: modelling using a medical care insurance claims database. PLoS One 13(3):e0195164. https://doi.org/10.1371/journal.pone.0195164

[8] Cheung S, Hamuro Y, Mahlich J, Nakayama M, Tsubota A (2018) Treatment pathways of Japanese prostate cancer patients—a retrospective transition analysis with administrative data. PLoS One 13(4):e0195789. https://doi.org/10.1371/journal.pone.0195789

[9] Endo A, Shiraishi A, Fushimi K, Murata K, Otomo Y (2018) Comparative effectiveness of elemental formula in the early enteral nutrition management of acute pancreatitis: a retrospective cohort study. Ann Intensive Care 8(1):1–8. https://doi.org/10.1186/s13613-018-0414-6

[10] Endo S, Ikeda N, Kondo T, et al (2019) Risk assessments for broncho-pleural fistula and respiratory failure after lung cancer surgery by National Clinical Database Japan. Gen Thorac Cardiovasc Surg 67:297–305. https://doi.org/10.1007/s11748-018-1022-y

[11] Etoh T, Honda M, Kumamaru H, et al (2018) Morbidity and mortality from a propensity score-matched, prospective cohort study of laparoscopic versus open total gastrectomy for gastric cancer: data from a nationwide web-based database. Surg Endosc Other Interv Tech 32(6):2766–2773. https://doi.org/10.1007/s00464-017-5976-0

[12] Fujimoto S, Nakayama T (2019) Effect of combination of pre- and postoperative pulmonary rehabilitation on onset of postoperative pneumonia: a retrospective cohort study based on data from the diagnosis procedure combination database in Japan. Int J Clin Oncol 24(2):211–221. https://doi.org/10.1007/s10147-018-1343-y

[13] Fujiogi M, Michihata N, Matsui H, Fushimi K, Yasunaga H, Fujishiro J (2019) Outcomes following laparoscopic versus open surgery for pediatric inguinal hernia repair: analysis using a national inpatient database in Japan. J Pediatr Surg 54(3):577–581

[14] Fujiogi M, Michihata N, Matsui H, Fushimi K, Yasunaga H, Fujishiro J (2018) Clinical features and practice patterns of gastroschisis: a retrospective analysis using a Japanese national inpatient database. Pediatr Surg Int 34(7):727–733. https://doi.org/10.1007/s00383-018-4277-6

[15] Fujiogi M, Michihata N, Matsui H, Fushimi K, Yasunaga H, Fujishiro J (2018) Postoperative small bowel obstruction following laparoscopic or open fundoplication in children: a retrospective analysis using a nationwide database. World J Surg 42(12):1–6. https://doi.org/10.1007/s00268-018-4735-2

[16] Fujita M, Sugiyama M, Sato Y, et al (2018) Hepatitis B virus reactivation in patients with rheumatoid arthritis: analysis of the National Database of Japan. J Viral Hepat 25:1312–1320. https://doi.org/10.1111/jvh.12933

[17] Gosho M (2018) Risk of hypoglycemia after concomitant use of antidiabetic, antihypertensive, and antihyperlipidemic medications: a database study. J Clin Pharmacol 58(10):1324–1331. https://doi.org/10.1002/jcph.1147

[18] Gouda M, Matsukawa M, Iijima H (2018) Associations between eating habits and glycemic control and obesity in Japanese workers with type 2 diabetes mellitus. Diabetes, Metab Syndr Obes Targets Ther 11:647–658. https://doi.org/10.2147/DMSO.S176749

[19] Hamamoto Y, Sakakibara N, Nagashima F, Kitagawa Y, Higashi T (2018) Treatment selection for esophageal cancer: evaluation from a nationwide database. Esophagus 15(2):109–14. https://doi.org/10.1007/s10388-018-0605-0

[20] Hamano H, Mitsui M, Zamami Y, et al (2019) Irinotecan-induced neutropenia is reduced by oral alkalization drugs: analysis using retrospective chart reviews and the spontaneous reporting database. Support Care Cancer 27:849–856. https://doi.org/10.1007/s00520-018-4367-y

[21] Harada M, Fujihara K, Osawa T, et al (2020) Association of treatment-achieved HbA1c with incidence of coronary artery disease and severe eye disease in diabetes patients. Diabetes Metab 46(4):331–334. https://doi.org/10.1016/j.diabet.2018.08.009

[22] Hiki N, Honda M, Etoh T, et al (2018) Higher incidence of pancreatic fistula in laparoscopic gastrectomy. Real-world evidence from a nationwide prospective cohort study. Gastric Cancer 21(1):162–170. https://doi.org/10.1007/s10120-017-0764-z

[23] Hirakata M, Yoshida S, Tanaka-Mizuno S, Kuwauchi A, Kawakami K (2018) Pregabalin prescription for neuropathic pain and fibromyalgia: a descriptive study using administrative database in Japan. Pain Res Manag 2018:2786151. https://doi.org/10.1155/2018/2786151

[24] Hiyama N, Sasabuchi Y, Jo T, et al (2018) The three peaks in age distribution of females with pneumothorax: a nationwide database study in Japan. Eur J Cardio Thoracic Surg 54:572–578. https://doi.org/10.1093/ejcts/ezy081

[25] Hosohata K, Inada A, Oyama S, Furushima D, Yamada H, Iwanaga K (2019) Surveillance of drugs that most frequently induce acute kidney injury: a pharmacovigilance approach. J Clin Pharm Ther 44:49–53. https://doi.org/10.1111/jcpt. 12748

[26] Hosomi K, Fujimoto M, Ushio K, Mao L, Kato J, Takada M (2018) An integrative approach using real-world data to identify alternative therapeutic uses of existing drugs. PLoS One. 13(10):e0204648. https://doi.org/10.1371/journal.pone.0204648

[27] Ikeda Y, Kubo T, Oda E, Abe M, Tokita S (2019) Retrospective analysis of medical costs and resource utilization for severe hypoglycemic events in patients with type 2 diabetes in Japan. J Diabetes Investig 10:857–865. https://doi.org/10. 1111/jdi.12959

[28] Imatoh T, Nishi T, Yasui M, et al (2018) Association between dipeptidyl peptidase-4 inhibitors and urinary tract infection in elderly patients: a retrospective cohort study. Pharmacoepidemiol Drug Saf 27(8):931–939. https://doi.org/10. 1002/pds.4560

[29] Ishikawa T, Obara T, Nishigori H, et al (2018) Antihypertensives prescribed for pregnant women in Japan: prevalence and timing determined from a database of health insurance claims. Pharmacoepidemiol Drug Saf 27(12):1325–1334. https:// doi.org/10.1002/pds.4654

[30] Ishimaru M, Matsui H, Ono S, Hagiwara Y, Morita K, Yasunaga H (2018) Preoperative oral care and effect on postoperative complications after major cancer surgery. Br J Surg 105:1688–1696. https://doi.org/10.1002/bjs.10915

[31] Iwatsuki M, Yamamoto H, Miyata H, et al (2019) Effect of hospital and surgeon volume on postoperative outcomes after distal gastrectomy for gastric cancer based on data from 145523 Japanese patients collected from a nationwide web-based data entry system. Gastric Cancer 22(1):190–201. https://doi.org/10.1007/s10120-018-0883-1

[32] Kakeji Y, Takahashi A, Udagawa H, et al (2018) Surgical outcomes in gastroenterological surgery in Japan: report of National Clinical database 2011–2016. Ann Gastroenterol Surg 2(1):37–54. https://doi.org/10.1002/ags3.12052

[33] Kanaji S, Takahashi A, Miyata H, et al (2019) Initial verification of data from a clinical database of gastroenterological surgery in Japan. Surg Today 49(4):328–333. https://doi.org/10.1007/s00595-018-1733-9

[34] Kawasaki R, Konta T, Nishida K (2018) Lipid-lowering medication is associated with decreased risk of diabetic retinopathy and the need for treatment in patients with type 2 diabetes: a real-world observational analysis of a health claims database. Diabetes Obes Metab 20(10):2351–2360. https://doi.org/10.1111/dom.13372

[35] Kawata M, Sasabuchi Y, Taketomi S, et al (2018) Annual trends in arthroscopic meniscus surgery: analysis of a national database in Japan. PLoS One 13(4):e0194854. https://doi.org/10.1371/journal.pone.0194854

[36] Kinoshita N, Morisaki N, Uda K, Kasai M, Horikoshi Y, Miyairi I (2019) Nationwide study of outpatient oral antimicrobial utilization patterns for children in Japan (2013–2016). J Infect Chemother 25:22–27. https://doi.org/10.1016/j.jiac. 2018.10.002

[37] Kinoshita T, Tanaka S, Inagaki M, Takeuchi M, Kawakami K (2018) Prescription pattern and trend of oral contraceptives in Japan: a descriptive study based on pharmacy claims data (2006–2014). Sex Reprod Healthc 17:50–55. https://doi.org/ 10.1016/j.srhc.2018.06.004

[38] Kobayashi H, Arai H (2018) Donepezil may reduce the risk of comorbidities in patients with Alzheimer's disease: a large-scale matched case–control analysis in Japan. Alzheimer's Dement Transl Res Clin Interv 4:130–136. https://doi.org/10. 1016/j.trci.2018.03.002

[39] Kodera Y, Yoshida K, Kumamaru H, et al (2019) Introducing laparoscopic total gastrectomy for gastric cancer in general practice: a retrospective cohort study based on a nationwide registry database in Japan. Gastric Cancer 22:202–213. https://doi. org/10.1007/s10120-018-0795-0

[40] Koizumi M, Ishimaru M, Matsui H, Fushimi K, Yamasoba T, Yasunaga H (2019) Tranexamic acid and post-tonsillectomy hemorrhage: propensity score and instrumental variable analyses. Eur Arch Oto-Rhino-Laryngology 276:249–254. https:// doi.org/10.1007/s00405-018-5192-0

[41] Kose E (2018) Adverse drug event profile associated with pregabalin among patients with and without cancer: analysis of a spontaneous reporting database. J Clin Pharm Ther 43(4):543–549. https://doi.org/10.1111/jcpt.12683

[42] Kubo S, Noda T, Myojin T, et al (2018) National database of health insurance claims and specific health checkups of Japan (NDB): outline and patient-matching technique. bioRxiv. https://doi.org/10.1101/280008

[43] Kunishima H, Ito K, Laurent T, Abe M (2018) Healthcare burden of recurrent Clostridioides difficile infection in Japan: a retrospective database study. J Infect Chemother 24(11):892–901. https://doi.org/10.1016/j.jiac.2018.07.020

[44] Matsui A, Morimoto M, Suzuki H, Laurent T, Fujimoto Y, Inagaki Y (2018) Recent trends in the practice of procedural sedation under local anesthesia for catheter ablation, gastrointestinal endoscopy, and endoscopic surgery in Japan: a retrospective database study in clinical practice from 2012 to 2015. Drugs Real World Outcomes 5(3):137–147. https://doi.org/10.1007/s40801-018-0136-y

[45] Momo K, Takagi A, Miyaji A, Koinuma M (2018) Assessment of statin-induced interstitial pneumonia in patients treated for hyperlipidemia using a health insurance claims database in Japan. Pulm Pharmacol Ther 50:88–92. https://doi.org/10.1016/ j.pupt.2018.04.003

[46] Naganuma M, Motooka Y, Sasaoka S, et al (2018) Analysis of adverse events of renal impairment related to platinum-based compounds using the Japanese Adverse

Drug Event Report database. SAGE Open Med 6:1–11. https://doi.org/10.1177/205 0312118772475

[47] Nakaharai K, Morita K, Jo T, Matsui H, Fushimi K, Yasunaga H (2018) Early prophylactic antibiotics for severe acute pancreatitis: a population-based cohort study using a nationwide database in Japan. J Infect Chemother 24(9):753–758. https://doi.org/10.1016/j.jiac.2018.05.009

[48] Nakajima M, Aso S, Matsui H, Fushimi K, Yasunaga H (2018) Clinical features and outcomes of tetanus: analysis using a National Inpatient Database in Japan. J Crit Care 44:388–391. https://doi.org/10.1016/j.jcrc.2017.12.025

[49] Noguchi Y, Ueno A, Otsubo M, et al (2018) A new search method using association rule mining for drug-drug interaction based on spontaneous report system. Front Pharmacol 9:197. https://doi.org/10.3389/fphar.2018.00197

[50] Noguchi Y, Ueno A, Otsubo M, et al (2018) A simple method for exploring adverse drug events in patients with different primary diseases using spontaneous reporting system. BMC Bioinformatics 19(1):1–7. https://doi.org/10.1186/s12859-018-2137-y

[51] Obinata D, Sugihara T, Yasunaga H, et al (2018) Tension-free vaginal mesh surgery versus laparoscopic sacrocolpopexy for pelvic organ prolapse: analysis of perioperative outcomes using a Japanese national inpatient database. Int J Urol 25(7):655–659. https://doi.org/10.1111/iju.13587

[52] Ogino M, Shiozawa A, Ota H, Okamoto S, Hiroi S (2018) Treatment and comorbidities of multiple sclerosis in an employed population in Japan: analysis of health claims data. Neurodegener Dis Manag 8:97–103. https://doi.org/10.2217/nmt-2017-0047

[53] Ohbe H, Jo T, Yamana H, Matsui H, Fushimi K, Yasunaga H (2018) Early enteral nutrition for cardiogenic or obstructive shock requiring venoarterial extracorporeal membrane oxygenation: a nationwide inpatient database study. Intensive Care Med 44(8):1258–1265. https://doi.org/10.1007/s00134-018-5319-1

[54] Okumura Y, Sugiyama N, Noda T (2018) Timely follow-up visits after psychiatric hospitalization and readmission in schizophrenia and bipolar disorder in Japan. Psychiatry Res 270:490–495. https://doi.org/10.1016/j.psychres.2018.10.020

[55] Okumura Y, Sugiyama N, Noda T, Sakata N (2018) Association of high psychiatrist staffing with prolonged hospitalization, follow-up visits, and readmission in acute psychiatric units: a retrospective cohort study using a nationwide claims database. Neuropsychiatr Dis Treat 14:893–902. https://doi.org/10.2147/NDT.S16 0176

[56] Okumura Y, Sugiyama N, Noda T, Tachimori H (2019) Psychiatric admissions and length of stay during fiscal years 2014 and 2015 in Japan: a retrospective cohort study using a nationwide claims database. J Epidemiol 29(8):288–294. https://doi.org/10.2188/jea.JE20180096

[57] Otaki Y, Watanabe T, Konta T, et al (2018) Effect of hypertension on aortic artery disease-related mortality 3.8-year nationwide community-based prospective cohort study. Circ J 82:2776–2782. https://doi.org/10.1253/circj.CJ-18-0721

[58] Noguchi Y, Katsuno H, Ueno A, et al (2018) Signals of gastroesophageal reflux disease caused by incretin-based drugs: a disproportionality analysis using the Japanese adverse drug event report database. J Pharm Heal Care Sci 4:15. https://doi.org/10.1186/s40780-018-0109-z

[59] Noguchi Y, Ueno A, Katsuno H, et al (2018) Analyses of non-benzodiazepine-induced adverse events and prognosis in elderly patients based on the Japanese adverse drug event report database. J Pharm Heal Care Sci 4(1):10. https://doi.org/10.1186/s40780-018-0106-2

[60] Oyama S, Hosohata K, Inada A, et al (2018) Drug-induced tubulointerstitial nephritis in a retrospective study using spontaneous reporting system database. Ther Clin Risk Manag 14:1599–1604. https://doi.org/10.2147/TCRM.S168696

[61] Ozaki T, Goto Y, Nishimura N, et al (2019) Effects of a public subsidization program for mumps vaccine on reducing the disease burden in Nagoya City, Japan. Jpn J Infect Dis 72(2):106–111. https://doi.org/10.7883/yoken.JJID.2018.276

[62] Ruzicka DJ, Imai K, Takahashi K, Naito T (2018) Comorbidities and the use of comedications in people living with HIV on antiretroviral therapy in Japan: a cross-sectional study using a hospital claims database. BMJ Open 8(6):1–11. https://doi.org/10.1136/bmjopen-2017-019985

[63] Ruzicka DJ, Tetsuka J, Fujimoto G, Kanto T (2018) Comorbidities and co-medications in populations with and without chronic hepatitis C virus infection in Japan between 2015 and 2016. BMC Infect Dis 18(1):1–11. https://doi.org/10.1186/s12879-018-3148-z

[64] Sado M, Ninomiya A, Shikimoto R, et al (2018) The estimated cost of dementia in Japan, the most aged society in the world. PLoS One 13(11):e0206508. https://doi.org/10.1371/journal.pone.0206508

[65] Saito A, Kumamaru H, Ono M, Miyata H, Motomura N (2018) Propensity-matched analysis of a side-clamp versus an anastomosis assist device in cases of isolated coronary artery bypass grafting. Eur J Cardiothorac Surg 54:889–895. https://doi.org/10.1093/ejcts/ezy177

[66] Sasabuchi Y, Matsui H, Lefor AK, Fushimi K, Yasunaga H (2018) Timing of surgery for hip fractures in the elderly: a retrospective cohort study. Injury 49(10):1848–1854. https://doi.org/10.1016/j.injury.2018.07.026

[67] Seki T, Takeuchi M, Miki R, Kawakami K (2019) Follow-up tests and outcomes for patients undergoing percutaneous coronary intervention: analysis of a Japanese administrative database. Heart Vessels 34(1):33–43. https://doi.org/10.1007/s00380-018-1224-3

[68] Shibata N, Kimura S, Hoshino T, Takeuchi M, Urushihara H (2018) Effectiveness of influenza vaccination for children in Japan: four-year observational study using a large-scale claims database. Vaccine 36(20):2809–2815. https://doi.org/10.1016/j. vaccine.2018.03.082

[69] Shinkawa H, Yasunaga H, Hasegawa K, et al (2018) Mortality and morbidity after hepatic resection in patients undergoing hemodialysis: analysis of a national inpatient database in Japan. Surg (United States) 163(6):1234–1237. https://doi.org/ 10.1016/j.surg.2017.12.033

[70] Sruamsiri R, Iwasaki K, Tang W, Mahlich J (2018) Persistence rates and medical costs of biological therapies for psoriasis treatment in Japan: a real-world data study using a claims database. BMC Dermatol 18(1):1–12. https://doi.org/10.1186/s12895-018-0074-0

[71] Sugihara T, Yasunaga H, Matsui H, Kamei J, Fujimura T, Kume H (2018) Regional clinical practice variation in urology: usage example of the open data of the National database of health insurance claims and specific health checkups of Japan. Int J Urol 26:303–305. https://doi.org/10.1111/iju.13840

[72] Sugisaki N, Iwakiri R, Tsuruoka N, et al (2018) A case–control study of the risk of upper gastrointestinal mucosal injuries in patients prescribed concurrent NSAIDs and antithrombotic drugs based on data from the Japanese national claims database of 13 million accumulated patients. J Gastroenterol 53(12):1253–1260. https://doi. org/10.1007/s00535-018-1483-x

[73] Sugiura K, Ojima T, Urano T, Kobayashi T (2018) The incidence and prognosis of thromboembolism associated with oral contraceptives: age-dependent difference in Japanese population. J Obstet Gynaecol Res 44(9):1766–1772. https://doi.org/10. 1111/jog.13706

[74] Suzuki S, Yasunaga H, Matsui H, Fushimi K, Yamasoba T (2018) Trend in otolaryngological surgeries in an era of super-aging: descriptive statistics using a Japanese inpatient database. Auris Nasus Larynx 45(6):1239–1244. https://doi.org/ 10.1016/j.anl.2018.03.001

[75] Suzuki S, Desai U, Strizek A, et al (2018) Characteristics, treatment patterns, and economic outcomes of patients initiating injectable medications for management of type 2 diabetes mellitus in Japan: results from a retrospective claims database analysis. Diabetes Ther 9(3):1125–1141. https://doi.org/10.1007/s13300-018-0407-3

[76] Tadokoro F, Morita K, Michihata N, Fushimi K, Yasunaga H (2018) Association between sugammadex and anaphylaxis in pediatric patients: a nested case–control study using a national inpatient database. Paediatr Anaesth 28(7):654–9. https://doi. org/10.1111/pan.13401

[77] Taguchi Y, Inoue Y, Kido T, Arai N (2018) Treatment costs and cost drivers among osteoporotic fracture patients in Japan: a retrospective database analysis. Arch Osteoporos 13(1):45. https://doi.org/10.1007/s11657-018-0456-2

[78] Takeuchi M, Ito S, Nakamura M, Kawakami K (2018) Changes in hemoglobin concentrations post-immunoglobulin therapy in patients with kawasaki disease: a population-based study using a claims database in Japan. Pediatr Drugs 20(6):585–591. https://doi.org/10.1007/s40272-018-0316-y

[79] Tanaka Y, Mizukami A, Kobayashi A, Ito C, Matsuki T (2018) Disease severity and economic burden in Japanese patients with systemic lupus erythematosus: a retrospective, observational study. Int J Rheum Dis 21(8):1609–1618. https://doi.org/10.1111/1756-185X.13363

[80] Taniguchi Y, Oichi T, Ohya J, et al (2018) In-hospital mortality and morbidity of pediatric scoliosis surgery in Japan. Med (United States) 97(14):2–5. https://doi.org/10.1097/MD.0000000000010277

[81] Tsuchiya A, Yamana H, Kawahara T, et al (2018) Tracheostomy and mortality in patients with severe burns: a nationwide observational study. Burns 44(8):1954–1961. https://doi.org/10.1016/j.burns.2018.06.012

[82] Uda K, Okubo Y, Shoji K, et al (2018) Trends of neuraminidase inhibitors use in children with influenza related respiratory infections. Pediatr Pulmonol 53(6):802–808. https://doi.org/10.1002/ppul.24021

[83] Uechi E, Okada M, Fushimi K (2018) Effect of plasma exchange on in-hospital mortality in patients with pulmonary hemorrhage secondary to antineutrophil cytoplasmic antibody-associated vasculitis: a propensity-matched analysis using a nationwide administrative database. PLoS One 13(4):1–12. https://doi.org/10.1371/journal.pone.0196009

[84] Umeda T, Hayashi A, Harada A, et al (2018) Low-density lipoprotein cholesterol goal attainment rates by initial statin monotherapy among patients with dyslipidemia and high cardiovascular risk in Japan—a retrospective database analysis. Circ J 82:1605–1613. https://doi.org/10.1253/circj.CJ-17-0971

[85] Usui T, Funagoshi M, Seto K, Ide K, Tanaka S, Kawakami K (2018) Persistence of and switches from teriparatide treatment among women and men with osteoporosis in the real world: a claims database analysis. Arch Osteoporos 13(1):54. https://doi.org/10.1007/s11657-018-0466-0

[86] Wake M, Onishi Y, Guelfucci F, et al (2018) Treatment patterns in hyperlipidaemia patients based on administrative claim databases in Japan. Atherosclerosis 272:145–152. https://doi.org/10.1016/j.atherosclerosis.2018.03.023

[87] Yamada S, Sato I, Kawakami K (2019) A descriptive epidemiological study on the treatment options for head and neck cancer: transition before and after approval of cetuximab. Pharmacoepidemiol Drug Saf 28(3):330–336. https://doi.org/10.1002/pds.4703

[88] Yonekura H, Ide K, Onishi Y, Nahara I, Takeda C, Kawakami K (2019) Preoperative echocardiography for patients with hip fractures undergoing surgery. Anesth Analg 128(2):213–220. https://doi.org/10.1213/ANE.0000000000003888

[89] Yoshida S, Ide K, Takeuchi M, Kawakami K (2018) Prenatal and early-life antibiotic use and risk of childhood asthma: a retrospective cohort study. Pediatr Allergy Immunol 29(5):490–495. https://doi.org/10.1111/pai.12902

[90] Yoshida T, Miyata H, Konno H, et al (2018) Risk assessment of morbidities after right hemicolectomy based on the National Clinical Database in Japan. Ann Gastroenterol Surg 2:220–230. https://doi.org/10.1002/ags3.12067

References

1. Kanaji S, Takahashi A, Miyata H et al (2019) Initial verification of data from a clinical database of gastroenterological surgery in Japan. Surg Today 49(4):328–333. https://doi.org/10.1007/s00595-018-1733-9
2. Yamana H, Moriwaki M, Horiguchi H, Kodan M, Fushimi K, Yasunaga H (2017) Validity of diagnoses, procedures, and laboratory data in Japanese administrative data. J Epidemiol 27(10):476–482. https://doi.org/10.1016/j.je.2016.09.009
3. Noguchi Y, Katsuno H, Ueno A et al (2018) Signals of gastroesophageal reflux disease caused by incretin-based drugs: a disproportionality analysis using the Japanese adverse drug event report database. J Pharm Heal Care Sci 4:15. https://doi.org/10.1186/s40780-018-0109-z
4. Hoffman KB, Dimbil M, Erdman CB, Tatonetti NP, Overstreet BM (2014) The weber effect and the united states food and drug administration's adverse event reporting system (FAERS): analysis of sixty-two drugs approved from 2006 to 2010. Drug Saf 37(4):283–294. https://doi.org/10.1007/s40264-014-0150-2
5. Yamada K, Itoh M, Fujimura Y et al (2019) The utilization and challenges of Japan's MID-NET® medical information database network in postmarketing drug safety assessments: a summary of pilot pharmacoepidemiological studies. Pharmacoepidemiol Drug Saf 28(5):601–608. https://doi.org/10.1002/pds.4777
6. von Elm E, Altman DG, Egger M, Pocock SJ, Gøtzsche PC, Vandenbroucke JP, for the STROBE Initiative (2007) The Strengthening the Reporting of Observational Studies in Epidemiology (STROBE) statement: guidelines for reporting observational studies. Ann Intern Med 147(8):573–578
7. Vandenbroucke JP, Elm E Von, Altman DG, Gøtzsche PC, Mulrow CD, Pocock SJ, Poole C, Schlesselman JJ, Egger M, for the STROBE Initiative (2007) Strengthening the Reporting of Observational Studies in Epidemiology (STROBE): explanation and elaboration. Epidemiology 18:805–835

8. Benchimol EI, Smeeth L, Guttmann A, Harron K, Moher D, Petersen I, Sørensen HT, von Elm E, Sinéad ML (2015) RECORD Working Committee. The REporting of studies Conducted using Observational Routinely-collected health Data (RECORD) statement. PLoS Med 12(10):475–487. https://doi.org/10.7507/1672-2531.201702009
9. Langan SM, Schmidt SA, Wing K et al (2018) The reporting of studies conducted using observational routinely collected health data statement for pharmacoepidemiology (RECORD-PE). BMJ 363:k3532. https://doi.org/10.1136/bmj.k3532

Ethical Issues of Data Secondary Use in Japan

Ethical, Legal, and Social Issues Pertaining to the Use of Real-World Health Data in Japan

Ryuichi Yamamoto

1 Introduction

Health care is based on medical science, and medical science cannot make progress without using human data. On the other hand, the right to privacy is an important consideration when using health data. While it is necessary to improve a person's health, treat their medical problem, or prevent it from worsening, it is also important to maintain the confidentiality of personal information. Some disorders can be inherited, the knowledge of which may cause one's child to be unfairly discriminated in marriage or employment.

Real-world health care thus deals with sensitive personal data. In Japan, workers dealing with health data, such as physicians and nurses, are bound by duty to respect the confidentiality of personal information. In fact, the Japanese criminal law has stipulated the health professional's duty of confidentiality for over a hundred years; violations are punishable by imprisonment, but they rarely occur. In 2005, the Act on Personal Information Protection (hereinafter referred to as "Act"), which comprehensively regulates the handling of personal information, came into effect. Various concerns surfaced as the Act was implemented, leading up to its revision in 2017. In the revised version of the Act, nearly all medical information is defined as sensitive information and stricter rules apply. It is clear that personal health data is protected not only by medical laws.

Prior to the Act, many health professionals abided by codes of ethics such as the Declaration of Helsinki and the Lisbon Declaration. The World Medical Association developed the Declaration of Helsinki as the code of ethics for medical research and the Declaration of Lisbon for the rights of patients. These codes prescribe not only how to handle patients' data but also patients' right to receive sufficient explanation,

R. Yamamoto (✉)
Medical Information System Development Center, Tokyo, Japan
e-mail: yamamoto@medis.or.jp

© The Author(s), under exclusive license to Springer Nature Singapore Pte Ltd. 2022   111
N. Nakashima (ed.), *Epidemiologic Research on Real-World Medical Data in Japan*,
SpringerBriefs for Data Scientists and Innovators 2,
https://doi.org/10.1007/978-981-19-1622-9_14

and that treatment should be provided with patients' consent. Physicians in most countries, including Japan, voluntarily comply with these codes of ethics. In 2016, the World Medical Association extended the Declaration of Helsinki by adding another set of ethical considerations regarding health databases and biobanks through the Declaration of Taipei.

2 Protecting Privacy Through the Act on Personal Information Protection

Privacy is a human right and a concept that is still evolving. Generally, the right to privacy implies that the confidentiality of personal information should be maintained and that it should be used only within the scope and purposes approved by the individual who provided the information. In theory, protecting this right is paramount. In practice, it becomes tricky.

In Japan, the Act on Personal Information safeguards privacy, but it is not intended to protect the rights of individuals per se. Rather, it prescribes the obligations of organizations that collect personal information. How much privacy each individual requires varies depending on the type of information collected and how much one perceives its importance. Some people do not mind if their address becomes known to others; for others, it creates serious problems. As it is difficult to resolve such individual differences through legislation, the Act originally imposed obligations on organizations who collect personal information. While the Act required the organizations to emphasize security, there was virtually no penalty, and all personally identifiable information was treated equally without differentiating it according to individual preferences. It was a legislation that did not correspond to the nation's views on the right to privacy, which inevitably causes complications. The Act was revised in 2017 to tackle such issues that arose. Effective penalties and special regulations were introduced. Some types of personal information were classified as sensitive data. Additionally, the Right to Request Disclosure was listed as a right.

3 Privacy in the Age of Big Data

In today's terms, privacy is the right of an individual to avoid an unfair disadvantage when the individual's information is handled in business. This concept had undergone changes depending on how personal information is used. Historically, as a human right that had been recognized since the latter half of the nineteenth century, privacy was defined as a right that prohibited an unwarranted disclosure of personal information if it had not been previously disseminated; this is because of a media that unilaterally commercialized personal information. The concept then evolved into the right to control one's own personal information in the second half of the twentieth century. In recent years, as the technology for handling large amounts of data had been advancing, and the progress of technology in the Internet of Things had led to a dramatic increase in health data, European Union members have discussed privacy as the right to be forgotten. Although the processing and utilization of these large amounts of newly generated health data is expected to contribute not only to maintaining and restoring people's health but also to improvement of health in a broader sense, the majority of health data consists of sensitive information, which increases the risk of privacy violation. The revision of the Act in 2017 introduced some countermeasures. However, further research remains necessary. The Act is to be reviewed every three years and will be revised again this year.

4 Ethical, Legal, and Social Implications

Although the Act is regularly reviewed, it is the legal responsibility of organizations to ensure the protection of the personal information collected. The Act emphasizes "consent" to ensure compliance to privacy as a human right. However, there is often a knowledge gap between the individual and the party who collects and/or utilizes the individual's personal information, in which the individual has difficulty understanding the explanation given and as such is ill-equipped to provide informed consent. This is especially true in the field of health care. A medical explanation, such as an explanation provided before a surgery, often carries major significance for maintaining and restoring the individual's health, but oftentimes consent is granted without adequate understanding. This becomes more complex when health information is used in big data: the possible outcomes of this are difficult to grasp compared to the possible outcomes of a surgery, which can often be understood even without details of the surgery. Thus, it is insufficient to merely legislate appropriate use of information and privacy protection. Ethical norms play a part, even after informed consent has been granted, for example. To increase the effectiveness of ethical norms, it is necessary to raise public awareness. Fortunately, discussions on ethical norms are held regularly on platforms such as the council sessions of the World Medical Association on the Declaration of Taipei. That said, in order to encourage appropriate use of real-world health data, it is necessary to integrate discussions with the legal system.

The Next-Generation Medical Infrastructure Law

Ryuichi Yamamoto

1 Japan's Progress in Introducing IT in Healthcare

Prior to 2000, Japan displayed globally top-ranking progress in introducing IT in healthcare. Insurance-claim computers (*rese-kon*) and medical computers (*iji-kon*), meant for creating itemized invoices for healthcare fees in Japan's universal healthcare system, had been introduced in the 1960s and spread relatively rapidly. The piecework-payment system of the time meant that creating these invoices involved the endless repetition of countless simple calculations, and medical institutions labored considerably on creating them. Although the computers of the 60s were rather powerless machines compared to the smartphones and advanced calculators we have today, any computer can handle the repetition of simple calculations with ease, and introducing them lightened the burden on medical institutions. Japan's healthcare system at the time had thus rapidly modernized, in the sense that it increased revenues by reducing the labor involved.

By the 80s, Japan's healthcare costs had skyrocketed, challenging society to streamline them without sacrificing the quality of healthcare. Systems to streamline bureaucracy, or "order-entry systems" or "ordering systems®," as they were called in Japan, began to be introduced predominantly at large hospitals, to eliminate administrative tasks with no direct relation to healthcare and tasks involving carrying around information written on paper, amongst other such tasks. The motivations for this were mostly economic, and since it achieved some measure of success in that regard, it spread relatively quickly among large hospitals. One feature of this paperwork-streamlining system is that it required information to be input into the system at its source. In other words, computers had now entered the site of medical treatment, and not only did this streamline paperwork, it also set in motion the development of the

R. Yamamoto (✉)
Medical Information System Development Center, Tokyo, Japan
e-mail: yamamoto@medis.or.jp

© The Author(s), under exclusive license to Springer Nature Singapore Pte Ltd. 2022　　115
N. Nakashima (ed.), *Epidemiologic Research on Real-World Medical Data in Japan*,
SpringerBriefs for Data Scientists and Innovators 2,
https://doi.org/10.1007/978-981-19-1622-9_15

electronic medical record (EMR) system, which helps healthcare practitioners and promotes patient-centered healthcare.

However, EMRs entail fewer economic incentives as compared to *rese-kon/iji-kon* systems and order-entry systems, which is why they did not spread as rapidly. Nonetheless, EMRs are gradually gaining ground as their merits become more widely known and have now been introduced at a majority of large hospitals and over 30% of clinics.

2 Operating Rules and Security

Thus far, whenever new health databases are planned, their operating rules for protecting privacy and guaranteeing security are discussed on an individual basis so that each one is safe and reliable. Even though the necessity of a unified operating standards or codes of ethics for health databases had long been debated, there is none currently. The protection of privacy is the sine qua non of a health database, which cannot coexist with privacy violations. While it is fundamental to obtain informed consent before collecting data, not all purposes for which the data may be used in specified in case of health databases. Often data collected is analyzed at a later stage after a certain amount of data is gathered. When collecting data at hospitals, it is common to not have any idea about whether or not a certain drug will have side effects. Obtaining strict informed consent is thus tricky to navigate; not all purposes of use can be specified at the time the data is collected.

In the absence of strict informed consent, what should be done to ensure that patients' rights are not violated or that certain medical institutions are not unfairly penalized? One option is to have a well-built legal system; another is to have a code of ethics that is widely accepted by people in different sectors, such as researchers, users of data, or data providers. The World Medical Association's Declaration of Taipei [3] is arguably a strong candidate for both. If providers and users of data write plans for research and studies that comply with a code of ethics, agreeing with the provisions of the legal system, it should be possible to more or less guarantee fair use. However, there is another important consideration: information security. This is the prevention of data from being stolen or leaked. If data gets stolen or leaked, one cannot know what will happen to it, making it impossible to control. Legal systems and codes of ethics are sufficient to respond to privacy concerns while the situation can still be controlled, but they are powerless after data theft or leakage, which makes information security very important.

Health databases do not function merely by having a computer with a storage device and building a database on it. As Fig. 1 shows, data must be collected at clinical sites and the use of that data must be allowed. Each of these sites has attack points, which must be attended to once they are identified. Addressing these points is hard work, and it is what arguably makes security such a difficult challenge. It

Security of Health Database

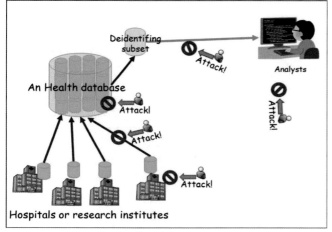

For using health database:
> Collecting data from various institutes,
> Re-organizing collected data,
> Setting database subsets for analyzing,
> (and so on)
There may be many attack points not only in Database but outside of it.

Fig. 1 Security in a health database

requires unceasing efforts to lower risks to acceptable levels while employing recommended guidelines such as the Ministry of Health, Labor, and Welfare's *Guidelines on Managing the Safety of Medical Information Systems.*

3 Japan's Legal System for Protecting Personal Information

This refers to the legal system mentioned in the previous chapter; the Act on the Protection of Personal Information [1] (hereafter, PPI Act), which was amended and then enacted in May 2017. In this chapter, I will discuss what was amended.

One feature of the amended PPI Act is that it introduced the idea of "individual identification codes" (hereafter, IIC) which are single pieces of information that can identify an individual. For example, passport numbers are IICs, but genetic information (such as complete genome/exome sequence data, single-nucleotide polymorphism [SNP] data, sequence data that includes 40 or more mutually-independent SNPs, and short tandem repeat [STR] data) can also be considered IICs if they are in a form that can be used to identify an individual. Because IICs can identify individuals on their own, they are impossible to anonymize. The fact that nearly all medical data was designated as "sensitive personal information" that must be handled with care is also thought to have been a significant impetus for these amendments. Sensitive personal information can only be collected upon obtaining consent, and if the person in question gave opt-out (implied) consent, its provision to third parties is prohibited.

These amendments prevent people's data from being provided to third parties unbeknownst to them by hospitals and by medical, caregiving, and health screening institutions. They have also eliminated the possibility that the results of commercial genetic analyses, which are prevalent these days, may be used for anything other than what was expressly consented to. In a sense, our legal system has made it easier to assuage the public's concerns.

However, there is more than just only using medical and health data for the sake of the patient. Development of medical science is based on clinical data, and the same is true of drug design and development of medical devices and technologies. The development of industries peripheral to medicine and caregiving should also be encouraged. Patients' medical and health data could be used for these purposes if patients' expressly consent to it; however, as mentioned earlier, it is rare that such explicit statements of purpose are given while data is collected for health databases. Additionally, even though there would be no risk of violation of privacy, Japan's amended legal system for protecting personal information may make it difficult to use this data. Something had to be done, and although it has not solved everything, the Next-Generation Medical Infrastructure Law was enacted as a first attempt to do so.

4   The Next-Generation Medical Infrastructure Law

The Act for Anonymized Health Information for Medical Research and Development (known as Next-Generation Medical Infrastructure Law [2] in short) was established with the intention of promoting the safe use of such data for public benefit in a broad sense of the term, such as for the development of medical science, drugs, medical devices, and medical technologies. This law is restricted to information that is necessary for medical research and the development of medical technologies as a matter of public benefit. As shown in Fig. 2, it stipulates that the government can authorize organizations that have the capability of anonymizing and providing data so that it will not violate the privacy of any individual whatsoever. It is designed as a separate law from the PPI Act, and it is noteworthy that this includes data from the deceased in its scope, which the PPI Act does not. Authorized organizations are called Authorized Anonymizing Health Data Organizations (hereafter, AAHDO). This permits a medical institution to provide data to an AAHDO from patients who gave opt-out (implied) consent (which, as mentioned above, is prohibited under the amended PPI Act) but only if that institution provides it to an AAHDO. However,

Outlines of Next Generation Medical Infrastructure Law

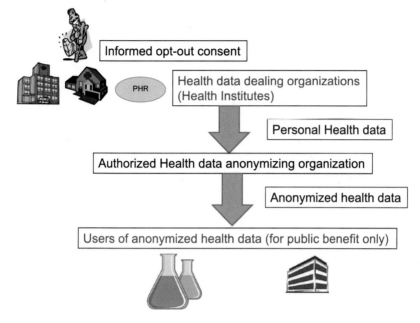

Fig. 2 Outlines of next generation medical infrastructure law

this is limited to cases where the institution's request is not be denied once they have notified the patient; this is sometimes called a "courtesy opt-out."

This means that AAHDOs are expected to use the data collected this way after carefully reviewing the purpose for its use to an appropriate standard, in order to ensure that that purpose can be achieved without causing trouble to any individual whatsoever or inflicting unfair discrimination on the medical institution that provided it. The Next-Generation Medical Infrastructure Law was put into effect in May 2018, and though it imposed strict information security conditions and required some time to authorize organizations, the first AAHDOs were authorized in December 2019. Several operational experiments are also underway towards creating a new certification.

However, this law is not without issues. Since the medical institutions provide AAHDOs with their data as individual information, it requires patients' consent. Although the law slightly reduces paperwork by accepting consent that is given implicitly via a notification to opt-out, the remaining paperwork is still troublesome. In practice, the data is anonymized at the AAHDO before it is used. Therefore, the government should make it possible to use the data without patients' consent, considering that it authorizes organizations capable of advanced anonymization and entrusts them with it. The provision of data by medical institutions to AAHDOs is voluntary, making it difficult to obtain comprehensive datasets.

The law is set to be reexamined every three years, and I look forward to future developments.

5 Conclusion

This has been a discussion on the present institutional state of Japan's health databases. While some challenges do remain, the Next-Generation Medical Infrastructure Law defines the first systemic provision that actively promotes the use of databases in health and medical fields based on the guarantee of safety. I look forward to future developments.

References

1. About the Act on the Protection of Personal Information (Personal Information Protection Committee (in Japanese). https://www.ppc.go.jp/personal/legal/. Accessed 10 Feb 2020
2. About the Next-Generation Medical Infrastructure Law (in Japanese). https://www8.cao.go.jp/iryou/index.html. Accessed 10 Feb 2020
3. WMA declaration Of Taipei on ethical considerations regarding health databases and biobanks. https://www.wma.net/policies-post/wma-declaration-of-taipei-onethical-considerations-regarding-health-databases-and-biobanks/. Accessed 10 Feb 2020

Printed in the United States
by Baker & Taylor Publisher Services